Columbia University

Contributions to Education

Teachers College Series

No. 86

AMS PRESS
NEW YORK

A Study of Organization and Method of the Course of Study in Agriculture in Secondary Schools

By

THEODORE HILDRETH EATON, PH.D.

TEACHERS COLLEGE, COLUMBIA UNIVERSITY
CONTRIBUTIONS TO EDUCATION, No. 86

Published by
Teachers College, Columbia University
NEW YORK CITY
1917

Library of Congress Cataloging in Publication Data

Eaton, Theodore Hildreth, 1877-
A study of organization and method of the course of study in agriculture in secondary schools.

Reprint of the 1917 ed., issued in series: Teachers College, Columbia University. Contributions to education, no. 86.
Originally presented as the author's thesis, Columbia.

1. Agriculture--Study and teaching (Secondary)--United States. I. Title. II. Series: Columbia University. Teachers College. Contributions to education, no. 86.
S533.E3 1972 630'.712'73 78-176740
ISBN 0-404-55086-X

Reprinted by Special Arrangement with Teachers College Press, New York, New York

From the edition of 1917, New York
First AMS edition published in 1972
Manufactured in the United States

AMS PRESS, INC.
NEW YORK, N. Y. 10003

CONTENTS

ORGANIZATION AND METHOD IN AGRICULTURE IN SECONDARY SCHOOLS

PURPOSE

The following study was undertaken with a view to learning something of the character of organization and method in the teaching of agriculture in secondary schools. By organization is meant the coordination of means to the attainment of the end professed. Included within this meaning is method, as referring more specifically to actual procedure in the inter-activity of teacher and pupil in the classroom, laboratory, shop, and outdoors.

Under these two heads particular attention has been given to the study of the curriculum as a revealing factor in the matter of organization and to the activity of the teacher as an instructive factor in the revelation of method. But more or less controlling modifiers of organization, such as qualification of pupils and of teachers, adequacy of equipment, location of school, type of school, and aid from the state, have not been ignored. Out of this coordination of means the attempt is made to evolve those factors which have dominated most conspicuously the organization and method used in the schools studied.

The point of view is taken at the start that agriculture is to be regarded as a 'mode of life', rather than as a mere activity of production; that agricultural education has to do with the preparation of youth for life in the country rather than merely with the 'getting things done on farms'. Hence a consideration only of means which have to do with successful production per unit of land or of labor is not sufficient. In a study of the curriculum, heed has been given to the provision of subjects in the schools for so-called agricultural students, other than those which deal with the profitable production of plants or animals. Agriculture in the schools should be a curriculum rather than a course.

Nevertheless, it is not uniformly so regarded. By the very nature of things emphasis is thrown in this study upon features that have the productive aspect. In the first place, the limitation of time necessitated a concentration upon aspects recognized distinctly as 'agricultural'; in the second place, the content of the so-called agricultural subjects is marked by its insistence upon the economic phases of country living, and this content, which makes up the great mass of the published work of the Department of Agriculture at Washington, of experiment station 'literature', of agricultural college bulletins, and of text and reference books, tends to appropriate to itself the designation 'agricultural'; in the third place, teachers of agriculture are for the most part men educated in the scientific aspects of production from farms to such an extent that, for instance, the study of English, or civics, under the heading of Agricultural Education, would seem to them a misfit; and finally, the stress laid upon the 'vocational' aspect by laws and officers regulating the distribution of state aid, has tended distinctly to the narrowing of the conception of agricultural education. Hence, if the dominant note of the study is struck in the consideration of the technical courses, that dominance is not far from representative of the attitude of those who control in secondary agricultural education. Inasmuch as the economic is the fundamental characteristic of the farmer's life, since his vocation dominates and modifies his life activity through the twelve months and the twenty-four hours, as with few other men, the distortion is less one of truth than of completeness.

Inasmuch as in these schools are met most of the important problems that are likely to meet the educator who is faced with the construction or reconstruction of the working plans for agriculture in the secondary schools, the study will have served a useful end if it enables the superintendent, principal, school board or teachers to find an answer to some of those problems. Out of the experience of the schools herein studied arise more or less definite adjustments of means to meet particular situations, more or less definite evasions of particular problems. In the final chapter a brief discussion of principles involved in making proper adjustment of various factors is attempted. As these in varied aspects appear in the organization and method of the schools studied, it is hoped they may prove of constructive value.

METHOD OF GATHERING DATA

The data used in the study were gathered by personal visits to fifty schools in sixteen states during the fall and winter of the year 1915–1916. As often as railroad schedules permitted, a whole day was given to each school. In very few instances was the visitor unable to arrive at the school before or near the opening of the day's session. In no case was the time spent at the school less than half a day, and in two cases the time so spent amounted to two days. Only once were two schools visited in a single day, and then the visitor was in attendance upon all 'agricultural' classes in either school. In two cases schools were not in session upon the arrival of the visitor, and in two the school day was given over wholly to examination. In the first instance the agricultural instructor, in the second the principal, gave his time to explanation of the work of the school in agriculture and to an exhibition of the school plant and of the outside work accomplished by students. In all cases visits were received in kindly fashion. In very few cases was there any apparent aversion to answering questions. With forty-five of the schools subsequent correspondence has been used for verification and amplification of data. All schools having catalogues or other printed matter concerning the work have furnished such freely, and further information has been gathered by correspondence with state departments of education and university and agricultural college departments of education. Personal conferences with Dr. Book, of Indiana, Professor A. V. Storm, of the University of Minnesota; Dr. A. D. Dean, Mr. L. S. Hawkins, and Mr. A. K. Getman, of the University of the State of New York; Dr. R. W. Stimson, of the Massachusetts Bureau of Education; and Mr. H. G. Whitcher, of the State Department of Education in New Hampshire, have proved of help. Particular acknowledgment is due the courtesy of these gentlemen in furnishing data, and to Prof. K. L. Hatch, of the University of Wisconsin; Mr. L. H. Dennis, of Pennsylvania; Prof. F. B. Jenks, of the University of Vermont, and President Andrew M. Soule, of the Georgia College of Agriculture, for material of value.

CLASSIFICATION

The schools visited have been somewhat arbitrarily divided into three groups according to type. The larger group, A, consists of schools which are in name and in fact, or in fact only, public high schools. All but one have entrance requirements calling for the completion of Grades VII, VIII, or IX, or an equivalent preparation; all have a four-year course of study, all maintain a length of school year between thirty-six and forty weeks and call for from fourteen and one-half to sixteen units of satisfactorily completed work for graduation. With them, for the sake of convenience, is included the one so-called 'junior' high school visited. In matters of curriculum, particularly in the 'academic subjects', there is much similarity to the common high school. The sub-classes in this grouping are:

A–1 City High Schools
A–2 Town or District High Schools
A-3 County High Schools
A–4 Approved Academies
A–5 Congressional District Schools
A–6 Junior High School

Perhaps a word should be said in explanation of the last three terms. An 'Approved Academy' is a familiar institution in the New England states. It is a secondary school supported in part at least by private, often church, interests, administered by a private board of trustees, but carrying certain courses of study approved by the state department of public instruction, that give it the standing of a high school with respect to tuitions, rebates, state aid, etc. Congressional district schools are familiar in the south. These are country-life high schools set up, maintained, and administered by the state. The 'Junior High School' is a public school maintaining a four-year course of study above the sixth grade, administered and supported by local authority, and duly approved by the state department of education.

Group B includes special agricultural schools more diverse in their organization, but alike in that they are specific institutions set up and maintained for the purpose of giving instruction in technology of agriculture, skill in agricultural processes, or both, and not for 'general education'. The sub-classes are self-explanatory:

B-1 State Agricultural Schools
B-2 County Agricultural Schools
B-3 Philanthropic Agricultural Schools

Group C includes but a single school. This is a private preparatory school, under church auspices, independent, with respect to support, administration and approval, of both local and state authority.

SELECTION OF SCHOOLS

Before making up a schedule of vis'ts, letters were written to state superintendents, state agents in agricutural education, and agricultural college professors of education, asking for lists of secondary schools in which agriculture is taught in their respective states, and for a checking on those lists of schools which are doing representative work in that particular. Checking of schools that were 'good' or 'representative' was made in nearly all cases, and was followed fairly closely in making out the schedule. In cases of doubt that school was visited in which it was known that the work had been for some time established.

SURROUNDINGS

In the matter of rural environment, fourteen of the thirty-nine schools of Group A are in villages or in the open country, six of the eleven schools of Groups B and C. But the advantage of the special schools in that particular is somewhat more than the figures indicate. Every one of the schools of Group B has on one or several sides, surroundings of open fields and farm land. On the other hand, only three of the schools of Group A may be regarded as unfavorably located with respect to access to the fields. From five to thirty minutes' walk will take students of the rest to farms. Perhaps, of the group, the most advantageously located in that respect are the congressional district schools. The table showing possession of land will throw further light on the matter. (See Equipment.)

No distinct correlation between organization and location is discernible, except as possession of land corresponds with more open surroundings. In the provision of outdoor work there is some relation between the factors, such provision being usual in the special schools of the open country, rare in the village or town high schools.

DESCRIPTION OF SCHOOLS

In order that the classification adopted may be more clear than otherwise it might be, a brief description is given of one school in each of the sub-groups. Though no single school, perhaps, may be selected as the true type of the group for which it stands, yet the attempt has been made to describe schools that lie between the extremes. Details in regard to classroom, laboratory, shop, and outdoor methods of teaching have been reserved for the section devoted to method.

A City High School

This is a four-year high school of thirty-six weeks to the school year, requiring for entrance completion of Grade VIII in the common schools. Agriculture is an elective subject by half units in the first and second years. The school has twenty teachers and a total enrolment of 467; girls 296, boys 171. Forty pupils have elected Agriculture, seven girls and thirty-three boys, of whom eleven are from farms. The average age of the boys in Agriculture is sixteen years. Up to 1914 Agriculture was merely a half-unit of Elementary Agriculture. This is the second year of the two-unit plan.

The school is housed in a fairly good modern brick building, on a small lot in a city of 10,000 population. The surrounding country is one of heavy clay farms underlain with bituminous coal. Most farmers have royalties from coal lands or work a part of the time in coal mines. Agriculture is backward, though there is some grain farming and the common Middlewestern combination of corn and hogs. Apple orchards are numerous and the trees grow thriftily, but systematic cultivation, spraying and grading of fruit are rare. The community is prosperous through the possession of coal lands, but largely indifferent to what is taught in the schools. The local board of trade has recently started a campaign to boom the agriculture of the region and has cooperated in agricultural club work and exhibitions with the extension division of the state college of agriculture. A number of the high school boys have won prizes in club work, but no regular cooperation with the agricultural instructor exists.

The principal is favorable to the work in agriculture, if not enthusiastic, and the same may be said of the attitude of the teacher. The one expresses a desire to 'make the work of real practical value to the student'; the other wishes 'to give a first-hand acquaintance with agriculture' sufficient to determine the predilection of the student for such a vocation.

The teacher of agriculture holds the degree of Ph.B. from a small sectarian college, and has gained what knowledge he has of the technology of agriculture through reading. He was, however, born and brought up on a farm. He has spent two summers in the study of education and botany at a university, and has for experience in teaching, one year in the grades, and six years as science teacher in the high school. Besides the ten periods in teaching in agriculture, he gives ten to Botany, and five to Zoology. His salary is $900 for the school year.

The so-called science course follows:

I		
English		5
Algebra		5
Latin		5
or		
German		5
Freehand Drawing		
Soils and Crops	1/2	5
Horticulture	2/2	

II		
English		5
Geometry		5
Latin		5
or		
German		5
Mechanical Drawing		2
Zoology		5
or		
Animal Husbandry	1/2	5
Farm Management	2/2	

III		
English		5
Advanced Algebra	1/2	5
Solid Geometry	2/2	
History		5
Botany		5

IV	
Physics	5
History	5
Elective	5
Elective	5

The method of classroom teaching in agriculture is an exceedingly slavish text-book recitation. One period of laboratory work has been planned for, but never given, there is no shop work, and outdoor work is conspicuous by its absence. During the recitation the teacher recommended attendance upon the local poultry show, but did not offer to go with the class. The class made a creditable attempt to render the text in their own words, but exhibited no particular interest in the subject in hand.

No adjustment of hours is made to meet the needs in agriculture, no attention is paid to season, and none to local needs, unless it appear in the injection of horticulture, which deals wholly with fruit growing, because gardening has been given in the grades. Some correlation is said to be made with agriculture in the botany classes, but the work witnessed in that subject was fully as stilted as the work in agriculture.

A Town High School, State Aided

This is a four-teacher high school, under a district superintendent. It is located in a far from prosperous town of about 2,000 population, in a rough and hilly country, where a Borden's milk station and a large dairy farm, producing certified milk, are the chief centers of industry. The attitude of the community and of the supervising officer is reported favorable to the work in agriculture, chiefly because it brings a substantial state aid to the school.

The total enrolment is fifty-four. In agriculture are ten girls and eleven boys. Eight of the boys are from farms. The work in agriculture is in its second year. Entrance requirements are an age of fourteen years or completion of Grade VIII in the elementary schools. No students have been admitted, however, on the age basis.

The teacher of agriculture is a graduate of the state college of agriculture, without pedagogical training, and with one year of experience in the teaching of high-school agriculture. He was brought up on a successful dairy farm, and has had a considerable experience in the handling of cattle. His attitude toward the work is one of sober enthusiasm. In agreement with the principal, he accepts the vocational aim in his work, but hardly looks to the turning out of ready-made farmers. In the school he has no other classes than those in agriculture. His salary is $1,100, with $200 additional for supervision through the summer of required home projects.

The class period is forty minutes, daily periods for each class in agriculture being double. In fall and spring three double periods per week are assigned to outdoor work for each class. As the weather grows colder, shop work is substituted for outdoor work in two of these double periods, and the other is given over to classwork. All agricultural classes are taught in the forenoon, and afternoons are given to extension and home project work. The teacher reports that this half-day arrangement of his time is looked upon unfavorably by other teachers and townspeople. They feel that he is underworked and overpaid, when he is absent from the schoolhouse during the afternoon session. Alternation of years is not yet necessary, but will come in with the third year.

The curriculum outlined for agricultural students is that suggested and approved by the State Department of Education.

GRADE IX		GRADE X	
English	5	English	5
Algebra	5	Geometry	5
Biology	5	Ancient History	5
Farm Shop and Poultry	10	Soils and Fertilizers and Farm Crops	10

GRADE XI		GRADE XII	
English	5	English	5
Physics	5	Chemistry	5
German	5	or	
Fruit Growing Animal Husbandry and Dairying	10	German	5
		United States History and Civics	5
		Farm Management and Agricultural Engineering	10

In order to meet the requirements of the state with respect to enrolment in agriculture for state aid, girls were urged into the classes at the start. In the second year's class there are ten girls and one boy. To meet this situation the teacher has been sane enough to depart from the required content and has made the work a study of fruits, vegetables, and flowers for the home. The establishment of a home economics department in the coming year is expected to remove most girls from agricultural classes.

No regular text-books are used in the class work in agriculture. Procedure is by topic assignment, in part general, in part to individuals. Reference books and bulletins are both freely used. Upon the generally assigned readings recitation is made in answer to prepared questions by the teacher. Upon the individual assignments students report to the class for criticism. With the completion of every large topic examination is held, and the slate wiped. Such examination was in process on the day of visit. The topic studied was poultry house construction with a view to preparation for the construction of small houses for home projects in poultry, which had already begun. The questions follow:

1. Give three materials used for a floor of a hen-house and give the chief advantages and disadvantages of each.
2. What are the three principal types of roof? Which is most common? Show different types by diagram.
3. Discuss the roosts in a hen-house, answering the following questions:

    a. How high should they be?
    b. What would be the objection to having them too high?
    c. How much roost space is required per hen?
    d. Should roosts be all on the same level? Why?
    e. How should they be fastened? Why?
    f. How should they be arranged in order to aid in keeping free from mites?

4. How much window light would be required for a hen-house twelve by eighteen feet? On which side of the house would you have the windows? Should they be so they can be opened? Should they be longitudinal or vertical in position? Why?
5. Give five different systems of ventilation and give the advantages and disadvantages of each?
6. Draw a floor plan of a house to hold 100 hens. Show the positions of the doors and windows, roosts and nests.

Only one boy completed the answers to all questions in the eighty minutes, but all the rest asked to be allowed to come in after school to finish. The attitude of the boys, particularly with regard to the floor plan called for, was one of serious interest.

In the second-year class, with ten girls, the topic for examination was varieties of apples. The questions follow:

1. Name four varieties of apples adapted for keeping late in the winter.
2. Name fifteen varieties of apples in order of time of ripening.
3. Give a brief history of one variety.
4. Name five varieties high in quality; three low in quality.
5. Give a list of six varieties suitable for a home orchard.
6. Draw a cross section of an apple showing the different parts.
7. What are the four leading commercial varieties?

8–9–10. Write opposite its number the name of each of the varieties on the table.

Several specimens of each of thirteen varieties were placed on the table for identification by appearance, feeling, odor, and taste. The slicing up and tasting of the apples appealed particularly to the class, which took the examination as a giggly guessing game.

As samples of the evident attempt of the teacher to adjust the parts of his work to the problem in hand the following may be cited. School opened in the season of fairs. The entering class judged poultry at the fairs at the same time that they studied breeds and varieties of poultry in the classroom. Following the fairs the boys selected or purchased the poultry to be used in their projects. Meantime the period for disposal of old hens and cockerels came on to dictate the next topic, the preparing and marketing of poultry. The laboratory work consisted of killing and picking and packing poultry. Shop work, altogether in wood, was given over wholly to making equipment for poultry houses.

In equipment the school is very scantily supplied. The school building is a wretched old plaster-cracked firetrap, on the third floor of which agriculture has been given separate quarters in three adjoining rooms, one of which is used for class work at long pine tables. The others are dumping places. In spite of the meagerness of the supply of apparatus, the supply of fruit, vegetables, and cereals for examination was unusually creditable. A large supply of flower pots cluttered one room. These were used by the girls for growing geranium and begonia.

Of land for agricultural purposes the school spared from fairly generous grounds only a few square rods for a hen yard, in which the teacher kept a small flock of pedigreed layers.

Adjustment of the agricultural course to local needs can hardly be considered close here. But the teacher reports an intention to emphasize dairying in particular. Adjustment to season and the supply of material at hand is marked. Recognition of group needs appears in the girls' agriculture. And there is some glimmering of recognition of the fact that the project may be a problem rather than an exercise.

A County High School

This is a four-year high school, including Grades VIII, IX, X, and XI, supported wholly by a state allowance of $3,000 per annum. Course, methods, and texts are dictated by the state department without regard to local conditions. Lack of supervision, however, leaves some freedom to an overworked principal.

The faculty consists of four teachers, including the principal and a practical mechanic for the teaching of courses in carpentry and mechanics, which have never yet been given. The enrolment is 140 boys and girls, all from the country. In this, the third year, the growth is marked as against ninety in the first year, and 111 in the second. Pupils board in the town or attend daily from their homes. The principal, one teacher, and a number of pupils are housed in unoccupied school rooms of a hideous but substantial building of brick.

The principal is a Bachelor of Science of a state university. He is farm born, and now owns a thirty-acre farm in the vicinity of the school. He has had ten years of teaching experience, and has 'read up' in agriculture as the occasion for teaching it has arisen. His whole day is given up to teaching, and his sixth week-day is given to supervision of rural schools. To meet this condition the high school is open on Saturdays and closed on Mondays. His salary is $1,500 on a twelve-month basis.

The school is located in a mining town wherein prosperity is confined to the few. The town is utterly raw and unattractive. Hogs, hens, and wandering mules dispute the right of way between puddles in the streets. The surrounding country is rough and mountainous, heavily wooded with pitch pine. The farms are sandy and poor, given over to corn, cotton, and razorback hogs. The attitude of the community is somewhat hostile to any attempt at secondary education. The school is still looked upon as an alien institution.

All students, boys and girls, are held to one or the other of the courses presented by the state. Girls take such agriculture as is given, but for some reason the boys have been exempted from domestic science. They take manual training from which the girls are freed.

COURSE A		COURSE B	
I		I	
English	5	English	5
Arithmetic 2/2	5	Arithmetic 2/2	5
Elementary Agriculture 1/2	5	Elementary Agriculture 1/2	5
General Science 2/2	5	Beginner's Latin 2/2	5
English History	3	English History	3
Geometrical Drawing	2	Geometrical Drawing	2
Manual Training	2	Manual Training	2
Cooking	2	Cooking	2
School Garden		School Garden	

II		II	
English	5	English	5
Algebra	5	Algebra	5
Horticulture 1/2	5		
(not given)			
Botany 2/2	5	Beginner's Latin	5
General History	5	General History	5
Domestic Science	2	Domestic Science	2
Woodworking	2	Woodworking	2
(not given)		(not given)	
Vocational Accounts	2	Vocational Accounts	2
School Garden		School Garden	
III		III	
English	5	English	5
Plain Geometry	5	Plain Geometry	5
Physics	5	Physics	5
Agriculture 2/2	5	Caesar, Books I–IV	5
(not given)			
Either Animal Husbandry, Dairy, Poultry, or Field Crops			
Home and Community Sanitation	2	Home and Community Sanitation	2
Farm Mechanics	2		
(not given)			
Domestic Arts	2	Domestic Arts	2
School Garden		School Garden	
IV		IV	
English	5	English	5
Solid Geometry 1/2	5	Solid Geometry 1/2	5
Algebra 2/2	5	Algebra 2/2	5
Chemistry	5	Chemistry	5
American History and Civics	5	American History and Civics	5
Rural Economics	3	Cicero, six Orations	5
(not given)			
School Garden		Latin Prose Composition	

The work in agriculture, of which the aim is said to be 'prevocational', is text-book recitation. Several of the boys have worked with the principal in growing about one-eighth acre of cotton on school land that they have cleared up. All have worked in the grading of an athletic field that is nearly ready for use. Into the school garden, of which the faculty is very proud, boys and girls have ventured on occasion. The garden is a heart-shaped enclosure of perhaps four square rods, largely in paths, but given over in part to bulbs and flowers. An industrious girl might use two or three hours a week in the care of it during fall and spring. It represents, however, what is accomplished under the caption 'School Garden' throughout the four years of the course outlined. In the schoolhouse are several ornamental plants in pots, which have suffered from the cold weather. These may, perhaps, be included in the school garden work.

Manual training is taught in a shop which has been fitted up in the basement by local carpenters, and is a faithful following of the prescribed process outline. The teacher is a woman graduate of a state normal school.

Correlation with country life in other subjects than agriculture may be noted in the vocational accounts, and sanitation. Otherwise there seems to be none.

No laboratory work is given in the sciences, because the school has no equipment. In books of reference it is poverty stricken. The only reference book in agriculture is a United States year book of ancient vintage, which had never been used. All that the school possesses is represented in two acres of raw land and a barely furnished building.

The struggle of teachers to carry out the rigid prescription of the state department, in the midst of poverty and unfriendly surroundings is an example of missionary spirit worthy of the highest praise.

A New England Academy

This school is under church backing and is administered by a private board of trustees. It has a modest endowment and some income from gifts. It serves as a local high school under state approval, with the right to tuition from townships served. The majority of the pupils are boarders, but about one-third attend by the day. The school has boys' and girls' dormitories, a gymnasium, and a fairly good frame school building.

The location is in a small village at the foot of a mountain in rough and beautiful country, well forested, with a few moderately prosperous dairy farms, and some prospect of successful orchard farming. Already the school has an excellent reputation as a preparatory school and the possession of it is a matter of local pride. The principal stands well with the community, and anything that he undertakes is fairly certain of the endorsement of the townspeople. This has been true of the work in agriculture, which is now in its fifth year.

The total enrolment of the school is ninety, fifty boys and forty girls. For entrance, an age of thirteen years and examination based on eighth-grade completion are required. The school has been growing slowly during the past few years, and the enrolment in agriculture has somewhat more than kept pace with the growth of the school as a whole. In 1911–1912 there were twelve boys in agriculture comprising fifteen per cent. of the total:

1912–13—18 boys in Agriculture, comprising 15 per cent. of the total
1913–14—25 boys in Agriculture, comprising 23 per cent. of the total
1914–15—20 boys in Agriculture, comprising 21 per cent. of the total
1915–16—28 boys in Agriculture, comprising 31 per cent. of the total

Agriculture is housed with physics and chemistry in the basement rooms of a separate building, and has a rather scanty laboratory equipment, consisting of a soils auger, tubes and balances, a few cereal and fertilizer samples, and a Babcock tester. There is a separate basement shop for woodwork and blacksmithing, in which the three forges smoke so badly as to require an adjustment of the hours for work to a time when the schoolrooms above are not occupied. There are benches and tools adequate for the needs of present classes.

The school has thirty-five acres of land, which can hardly be regarded as a farm. There is half an acre of garden, cared for during the growing season by hired 'help', which is of some use for work in agriculture, a small orchard of apple trees, the pruning and spraying of which is done by students in agriculture; a five-acre plantation of young white pines set out by students; and a small herd of grade dairy cattle, which furnishes milk for the boarding pupils. The herd has been tested by students and used for judging purposes.

The principal is a Master of Arts and a Bachelor of Divinity, who looks upon agriculture as a liberalizing rather than a vocational subject, and has encouraged boys whose purpose is not definitely preparation for college to elect that work. The teacher of agriculture is a young man, city born, whose farm experience has consisted in one summer's work on a farm. He is a graduate of an agricultural college, has had a summer's course in pedagogy, and one year's experience as a high school teacher of agriculture. With respect to the aim of his work, he is a little uncertain, but desires to make the course practical, even if it does not turn out farmers. The shop teacher, who also teaches the forestry course, is farm born, but has no college training. He is a practical workman of the best type, and an enthusiastic teacher. Five other teachers are enrolled in the faculty, among them a teacher of domestic science, who carries a course in agriculture for girls.

The boys in agriculture average in age seventeen years and two months; twenty are from farms, eight from towns. Their attitude in the agricultural classes was largely one of indifference, but in the active work of the shop they were taking hold with a will.

The approved curriculum for a four-years' course of thirty-seven weeks per year is as follows:

GRADE IX		GRADE X	
Soils and Crops	5	Farm Crops and Animal Husbandry	5
Art	2	Art	2
English	5	English	5
Algebra	5	Geometry	5
Farm Carpentry	5	Forge Work	5

GRADE XI		GRADE XII	
Dairying Horticulture	5	Horticulture Farm Economics Rural Sociology	7
English	5	English	4
Forestry	5	United States History and Civics	4
Physics	6–7	Chemistry	5

In the approved curriculum in domestic science for girls is a single unit of agriculture, in the first year consisting of home gardening, home decoration, and floriculture, as a direct adjustment to the needs of girls. The class is taught by a woman and is largely text-book work. In fact, the work in the agricultural classes is dominantly recitation from assigned lessons in text-books. The question and answer method used reveals no skill on the part of the teacher, and allows of little initiative on the part of pupils. With text open, and scanning the pages, the teacher confined himself to snap questions, many of them leading, designed to show whether or no the pupils had read the assigned lesson. The result was a rather uninteresting confirmation of the fact that most of them had.

The outdoor work for the day was a lesson on weeds, of which the garden furnished an ample assortment. Each boy was provided with a mimeographed outline of the points to be noted. Most of the points had to do with botanical distinctions, such as arrangement of leaves on the stem, and nature of inflorescence. As the leaves were withered and the flowers gone to seed, the notes were filled in by guesswork as the boys lay at ease under the trees. When it was found that only five weeds were recognized either by boys or teacher, all others were cast aside and the note-taking confined to the five. No point definitely leading to method of control of the weed entered into consideration at all, despite the plan of the lesson in advance.

In shop work the teaching was much more satisfactory. Everybody was busy with a particular project of his own, following his own working drawings. The teacher had little to say. A few times boys came to him with questions as to procedure, which he answered carefully either in words or by demonstration. At the end of the period every boy brought his work up for criticism, which was frank and fair.

No farm work is required or provided, though the shop boys have done some repair jobs and construction work, notably, a concrete retaining wall of very creditable workmanship. Home and school projects are prospective only. The plan is to provide home work for day pupils and individual plots for boarders, who will have to hire others to carry on their garden work during the summer.

The school hours are sixty minutes in length, twenty minutes of which is given to supervised study, if the teacher is so disposed. There are no double periods in agricultural courses or in shop work, and the shop instructor reports himself handicapped by the arrangement. Since the school receives no state aid, it is felt to be necessary for the agriculture teacher to give part of his time to other work. Accordingly, he has charge of the classes in physics and chemistry. In neither course is any attempt at correlation made with the work in agriculture. The reason given is that "the college entrance requirements stand in the way of it." In order that the teacher may have time for the additional work in science, an alternation of yearly topics in agriculture is practised. Thus Freshmen, this year, take soils and crops along with Sophomores; Juniors and Seniors take dairying and horticulture together. Next year the work of the second and fourth years in agriculture will be given.

In this school the more notable adjustments appear:

1. As to local needs, in the introduction of forestry as a full unit and the emphasis upon dairying in the animal husbandry course.

2. As to needs in the line of progress, the emphasis upon fruit growing in horticulture.

3. As to seasonal demands, the provision of a spring to fall sequence for the completed cycle of growth in crops and horticulture.

4. As to group distinctions of pupils, in the provision of a special unit of agriculture for girls.

A Congressional District School

A school established by the state in one of the congressional districts of a Southern state, receiving $10,000 per year from the state for maintenance. Administratively, the school is a department of the agricultural college. The curriculum is uniform with that of the other congressional district schools and is laid down in annual conferences of the school principals with the authorities of the agricultural college and the professor of secondary education at the university. Occasional visits by the latter officer constitute the supervision.

The school is located in the open country, three miles from the nearest village and railroad station. The surrounding country is rough or rolling, much of it in woods. The soil is rather light and poor and the farming backward. Cotton, corn, and hogs are the chief products, but there is some peach orcharding. After eight years of growth the school has succeeded in changing the attitude of local hostility to one of tolerance that bids fair to become friendly.

The school owns and operates a farm of 250 acres, 100 acres in crops, cotton, and forage, and fifty in pasture. The state of tillage is good and the crops fair. A small orchard and demonstration plots are located near the main buildings. Farm buildings are decrepit and badly arranged, but a new concrete silo marks the start for a new set of better ones. A large gasoline engine in a concrete pump house delivers water from a 'run' to the school buildings. Another furnishes power for sawing wood, a dynamo, and for the small shop. Of horse machinery there is a reasonable supply in fair condition. The poultry plant has an incubator and brooders.

An excellent flock of Rhode Island Red hens marks the most progressive selection in live stock. A herd of rather inferior grade Jerseys supplies milk. A score of hogs and two teams of mules complete the quota of live stock.

The school buildings are of brick. A main recitation building, dining hall, and girls' dormitory built on the characteristic plan of these schools, under contract, show the characteristic bad workmanship noted in other schools. The roof leaks, the walls are damp, and the plaster is falling away, the floors are uneven. In contrast is the boys' dormitory, constructed by student labor after a fire had destroyed the original building. Though most of the brick used in the walls is that left from the fire, even the appearance of the building is more substantial than that of the others. The building will be usable when the others have fallen to ruin.

In this school the handicap of debt is one that is frequent under the present system. New construction, equipment, and repairs, must be undertaken from the maintenance allowance. Bad original construction, for the school is in the eighth year only, has entailed a heavy burden for repairs. Under a new law some relief is looked for. Division among the schools of inspection fees on fertilizers and agricultural products is expected to bring the annual income from the state to $15,000.

Eight teachers, including the principal, make up the faculty. The enrolment is 215, 144 boys and seventy-one girls. All boys take agriculture, and ninety per cent. of them are from farms. Entrance requirements stipulate completion of Grade VII, but enforcement is not strict. Twenty-four pupils have been admitted on approval by the principal. The average age of boys is eighteen.

The principal holds his A.B. from the state university. He was born and brought up on a farm and has had ten years of teaching experience. He conceives the aim of his school to be preparation for country living. He is vigorous and enthusiastic, a man of his hands as well as an unusual man in the classroom. But, though the management of farm as well as school is in his hands, the teaching of agriculture is not.

Science and agriculture are taught by a young man recently graduated from the school itself. He has had one year of experience in a one-room school and is in his second year in his present position. His aim is to relate the productive science of agriculture to the unity of which it is a part. Despite a farm bringing up and his training, he is handicapped by a limited knowledge of his subject. His salary is $540 for the thirty-six weeks of the school year; that of the principal, $1,500 and board for himself and family throughout the year.

The course of study is that agreed upon at the conference above mentioned. No laboratory work, however, is actually given, because the school has no equipment for it. Otherwise, the outline is followed:

I		
English	5	
Arithmetic	5	
United States and State History	4	
Advanced Geography	4	
Spelling and Writing	1	
General Agriculture	4	
Poultry	1	
Free Hand Drawing 1/2	1	(3)
Woodwork 2/2	1	(3)
Domestic Arts	2	(3)

II		
English	5	
Arithmetic	2	
Algebra	3	
European History	3	
Biology and Sanitation	5	
Spelling and Writing	1	
Breeds	2	
Dairy	1	
Stock Judging	1	(3)
Farm Crops	2	
Forge Work	1	(3)
Domestic Science and Art	3	(3)

III		
English	5	
Algebra, 7 months	5	
Geometry, 2 months	5	
European History	3	
Physics	3	
Teacher Training	3 *	
Feeds and Feeding 1/2	3	
Feeds and Feeding 2/2	2	
Elementary Horticulture 1/2	2	
Elementary Horticulture 2/2	3	
Feeds Lab. (not given)		
Horticulture Lab. (not given)		
Mechanical Drawing	1	(3)
Domestic Science and Domestic Art	3	(3)

IV		
English	5	
Geometry	3	
Advanced Civics	3	
Chemistry	3	
Teacher Training	2 *	
Soils 1/2	3	
Soils 2/2	2	
Fertilizers 1/2	2	
Fertilizers 2/2	3	
Elementary Surveying	1	(3)
Domestic Science and Domestic Art	3	(3)

* Substituted for work in Agriculture or Domestic Science.

The classroom work in agriculture is a rather dull text recitation as faithful to the text as may be. In the crowded room where some students were obliged to sit on the floor near a hot stove, several fell asleep, and lost nothing by it.

The shop work emphasizes mathematics and drawing and follows a process manual. But the boys have accomplished a good deal of repair and construction work. All repairs are made by students, and, in addition to the dormitory mentioned, they have erected a large concrete silo, a hen house with incubator room, the shop itself, and the pumphouse. Students are chosen to run the engines and dynamo.

Outdoor work to the extent of thirty hours per month is required of all students. They have charge of dormitories and school buildings and of all work but cooking in the dining hall. They work by assignment under supervision on alternate half days; the upper classes on one, the lower on the other. The tasks assigned arise out of farm and school needs, according to season. All work above the required thirty hours is paid for at from five to ten cents an hour, according to the work and the ability of the student. Some students have paid all expenses by means of such work.

Students may have for their own use one-tenth acre of land of which the product shall be theirs. Prizes are offered for most economical production from these plots. Over summer the principal takes care of the plots, charging regular rates against the crop for horse and man labor. Experimental plots, according to the design of the agricultural experiment station, are under direction of the teacher in agriculture and cared for by students in his classes.

No close correlation between classroom and outdoor work occurs, except as accident may bring it about. In chemistry, the attempt is made to tie up with work in agriculture, and the survey work is directly applied to the farm, but in other particulars, no correlation with productive agriculture is apparent.

Nine volumes of reference constitute the agricultural library.

The earnestness and interest of students seemed unusual. It was less evident in the productive agriculture than anywhere else.

Junior High School

This is a new school, housed in an excellent new building, but already, in its first year, overcrowded. All grades from I to XI are housed in the building under five teachers, two of whom give all their time to the 'Junior High School'. In the state the elementary school runs through nine grades, but the division line in this school is drawn at the close of grade VI. Above this grade the school is divided into four groups at present made up as follows: I, Grade VII and those of the lower standing in Grade VIII; II, Upper Grade VIII and Grade IX; III, Grade X; IV, Grade XI. I and II are taught in a room together; III and IV in another room on the second floor. In the basement is a room neatly equipped for domestic science and another neatly, but scantily, equipped for agriculture.

The school building stands in a large lot on the edge of a small but neat and prosperous village. Behind it rears a beautiful mountain, wooded nearly to the peak. Hay and dairy farms are scattered along the river and on the hills, but lumber is still the principal crop. The school is a matter of pride with the community and the introduction of agriculture and domestic science under state approval has been favorably received.

The principal is a young man just graduated from the state university, where he took his B.S. from the agricultural college. In his last year he took courses in pedagogy and practice teaching in rural schools and in agriculture in the city high school. He is enthusiastic and earnest. His expressed aim in agriculture is to make the work 'practical'. He teaches mathematics, geography, hygiene, and agriculture, having only four periods a week free from teaching. His salary is $900 for the academic year of thirty-six weeks.

The total enrolment is 148, sixty-four in the Junior High, thirty-four girls and thirty boys. All boys take agriculture, are from farms, and every one has a garden and can milk.

School periods are thirty minutes, with ten minutes for supervised study. Agriculture comes in the last periods of the afternoon to allow of outdoor work, which had, in the first five weeks, consisted in one excursion for the study of soils origins.

The curriculum follows:

I		II	
Arithmetic	5	Arithmetic	5
English	5	English	5
Writing and Spelling	5	Writing and Spelling	5
Geography	5	Reading	5
Reading	5	Geography and Hygiene	5

III		IV	
Algebra	5	Geometry	5
English	5	English	5
Writing and Spelling	5	Writing and Spelling	5
United States History	5	United States History and Civics	5
Agriculture	5	Agriculture	5
or		or	
Domestic Science	5	Domestic Science	5
or		or	
Latin	5	Latin	5

One period a week in agriculture is given to lecture and two double periods to text-book recitation or laboratory demonstration by the teacher. The boys at present are divided into two groups of twenty-one and nine, the smaller group being the more advanced. Both groups study the same topics in the same text, but the smaller group proceeds more rapidly. There is no shop work and no projects are in use or in prospect. Outdoor work is at a minimum.

The only adjustment noted in agriculture was the selection of variation and heredity as the first topic, to convince pupils that agriculture is no 'snap course'. Five weeks of the mountain autumn season had been given to this effort. The class appeared convinced. The interest in agriculture was very faint indeed. On the other hand, the interest in arithmetic was very marked, and the class work remarkable for the spontaneity and initiative of pupils. No set text was used, and to each pupil was assigned the task of making a text of his own from materials of his own home experience. The correlation with farm life appeared very close. In no other subject, however, appeared any attempt to teach in terms of country life.

The maintenance of interest in individual problems even in this one subject promises to be a considerable task for a teacher so heavily overloaded with work, but the start made is very encouraging.

A State Agricultural School

This is one of the special schools of agriculture and home economics provided by the state, having an aim primarily vocational. Support of the school is through legislative appropriation.

The school offers a regular two-year course, requiring for entrance an age of sixteen years, and completion of Grade VIII, and short winter courses for which only the age requirement is necessary. Pupils are for the most part boarders, who find rooms and meals in the village, though some go back and forth daily to their homes. Total enrolment is 193, boys 120, girls seventy-three. In the two-year course of agriculture are fifty boys, thirty-four of whom are from farms, in the short course twenty-six farm boys. The students are somewhat more mature than those found in the high schools, and show a greater variation in school preparation, as the following figures on age and preparation make plain:

Preparation	*Class 1916 (boys)*	*Class 1917 (boys)*
Some college work	1	0
Completed high school	10	5
Three years high school	6	6
Two years high school	18	12
One year high school	19	13
Completed grade VIII	15	13
Completed grade VII	1	0

Age at Entrance	*Class 1916 (boys)*	*Class 1917 (boys)*
16 years	9	0
17 years	10	7
18 years	17	5
19 years	11	8
20 years	10	7
21 years	8	10
22 years	2	7
23 years	2	1
24 years	2	3
25 years	0	1
26 years	1	1
29 years	0	1

Average age, 19.66 years. 20.8 years.

On the basis of preparation boys have been separated for instruction into three groups: A, 3 years or more of high school; B, 1 and 2 years of high school; C, completion of Grades VII or VIII.

The school is now in its sixth year and shows a fairly consistent growth as may be noted by the enrolment of boys in agriculture:

YEAR	NUMBER ENROLLED IN AGRICULTURE
1910–11	32
1911–12	34
1912–13	47
1913–14	55
1914–15	103
1915–16	78

Estimated average percentage of farm boys eighty-five per cent. The length of the year is thirty-four weeks. Periods of class work are forty minutes, with double periods for laboratory and shop work.

The faculty consists of twelve members. The principal is a Bachelor of Science, of a state university, who was for four years head of the department of agriculture in a large preparatory school, and has been principal here since the start. He was born in the city, but moved to a farm when a baby and was a farm boy till he entered college. His salary is $2,700 and house on a twelve-month basis.

In charge of academic work for the boys is a Master of Arts, who before entering college was a mechanic in a small country town. He studied education during two years of college, and is now in his sixth year as teacher in this school. His salary is $1,500 on a 12-month basis.

The teacher of agronomy and farm mechanics holds his Master's degree in agriculture, from an agricultural college. He was farm born and raised. For two years he was instructor in an agricultural college, and has taught in his present position since the school started. Salary $1,500, twelve months.

The teacher of animal husbandry and dairying holds his M.S. from an agricultural college. He was farm born and brought up, and has served a year and a half as instructor in an agricultural college. He has held his present position for six years. Salary $1,500, twelve months.

For teaching horticulture and botany is employed a young man, farm born and reared, who has his B.S. from an agricultural college. For a year after graduation he managed a large truck farm. He is in his second year as teacher. Salary $1,500, twelve months.

The poultry work instructor is a B.S. in agriculture, not farm born, but of several years' experience in farming. In college he took work in education. Teaching experience, three years, salary $1,500, twelve months.

For veterinary science a practising veterinarian has been employed on a part time basis since the school started. Stipend, $1,000.

The instructor in chemistry, English, and athletics is a city-born youth, holding his B.S., not in agriculture. He has taught here for four years. Salary $1,100, twelve months.

In shop work a practical mechanic has been employed since the school started. Salary $1,000, twelve months.

In agronomy and horticulture a graduate of the school acts as assistant at a salary of $900.

The regular course, which is differentiated into special lines in the second year, is outlined as follows:

JUNIOR YEAR

Fall Term		*Spring Term*	
English	3	English	3
Farm Arithmetic	3	Bookkeeping	2
Botany	4	Poultry	2
Chemistry	3	Chemistry	3
Physiology	3	Breeds	2
Soils and Manuring	4	Farm Crops	3
Stock Judging	1	Stock Judging	1
Shop	2	Shop	2
Drawing	1	Drawing	1
		Elementary Horticulture	2
		Dairying	2

SENIOR YEAR

Fall Term

Required of all Courses

English	3	United States History and Civics	3	Soil Fertility	2

SENIOR YEAR

Fall Term

Course in General Agriculture		*Course in Animal Husbandry and Dairying*		*Course in Poultry*		*Course in Horticulture*	
Stock Feeding	3	Stock Feeding	3	Poultry Management	3	Fruit Growing	4
Agricultural Engineering	2	Dairy Products	4	Marketing Poultry	2	Insect Pests	2
Farm Crops	2	Stock Judging	1	Dairying or Fruit Growing	3	Floriculture	2
Forging and Repairs	2	Veterinary Science	2	Elective	5	Elective	5
Drawing	1	Elective	3				
Elective	3						

Spring Term

Required of all courses

English	3	United States History and Civics	3	Farm Management	3

Course in General Agriculture		*Course in Animal Husbandry and Dairying*		*Course in Poultry*		*Course in Horticulture*	
Soil Fertility	2	Animal Management	3	Poultry Management	3	Orchard Practice	2
Animal Management	3	The Horse	1	Poultry Problems	1	Plant Diseases	2
Farm Machinery	2	Dairy Products	2	Incubation and Brooding	2	Sprays and Spraying	2
Shop	2	Stock Judging	1	Dairy Products or Orchard Practice	2	Market Gardening	2
Elective	3	Veterinary Science	2	Elective	4	Elective	4
		Elective	3				

Classroom work savors in method of the agricultural college. Lectures predominate, but text-book recitation and discussion of topics form a part of the work. In one class visited the lecture was a virtual dictation from typewritten outlines. Students were occupied mostly with the business of trying to keep up with the lecturer. At the end of the period the instructor announced that the contents of the lecture were to be found in a state experiment station bulletin, which might be had for the asking. As a drill in writing at top speed the lecture was a success. Yet this lecture, like the others heard, showed an excellent preparation on the part of the teacher and a definite attempt at application to concrete cases of established principles of control.

The laboratory work, as seen in the soils class, was a performance of experiments, individually, in accordance with a printed manual of directions. The instructor moved quietly among the boys, giving directions as to what should be observed and noted in the record books. The dairy laboratory class was at work pasteurizing milk, testing samples, and churning butter. Though the laboratory was well equipped, the class was too large for effective work. Six boys kept busy at the assignments while the other twenty-six stood round waiting for the period to expire.

Of farm work no regular requirement is made. Outside work is in charge of the instructor whose particular subject is involved. In horticulture the attempt at tying up the outdoor work with the indoor class work was marked. For instance, while the class lecture dealt with propagation by cuttings, boys were assigned to starting of cuttings in the greenhouse.

No regular distinction is made in outdoor work between city and farm boys, but the principal has given directions that with new students tasks involving primary skills, such as harnessing and milking, shall be performed by city boys, whereas weighing and mixing of rations, preparation of sprays, and like work not performed by boys on ordinary farms shall be given first to country lads. It is his desire that students shall be given opportunity at the school to acquire those skills which they are unlikely to learn on the home farm. Some definite grouping and rotation in processes through regularly required farm work seems desirable to this end. Some observation work is done on neighboring farms, but most of the practice is on the school farm.

Shop work is by process sequence with little or no attention to the useful individual project.

Correlation between agricultural and other subjects occurs in arithmetic, botany, drawing, and English.

Recurrent treatment appears in the sequence of specialized courses after the general course of the first year.

Classrooms are in the buildings of the former county court, but there is a new demonstration building with a generous judging pavilion, and a well-kept greenhouse for work in botany and horticulture. The farm equipment is excellent, without being over-expensive and elaborate. The shop is large, well-lighted, and amply equipped. The school farm is of 200 acres of good land, rolling, and mostly in crops. A young orchard of seven acres, two acres of small fruit, and a two-acre garden are close by the school buildings. Recently a tract of 175 acres of forest land, some miles from the school, has been acquired.

In live stock the farm is fairly furnished. The herd consists of forty-five pure bred and grade dairy cattle, housed in an up-to-date and business-like barn. There are two teams of draft horses, and one light team, pure bred Yorkshire, Berkshire, and Cheshire hogs, small flocks of Shropshire and Rambouillet sheep, and an excellent poultry plant, with flocks of pure bred fowl. On the whole, there is little on the farm that might not be maintained on a good business farm of like size and soil. A student might leave such a school with the hope of one day matching its conveniences on his own farm, whereas he is likely to cast from his mind as hopeless the elaborate and expensive improvements to be found at many of the colleges of agriculture. The library has 189 volumes on agriculture.

The community served is the state. No special adjustments to local needs appear. Seasonal adjustment appears in the term sequence of subjects.

A County Agricultural School

The farm and plant of this special vocational school of Agriculture and Domestic Economy were provided by the county four years ago. Toward maintenance the county contributes annually $3,000, the state $6,000. The farm of 128 acres is a good one, and though managed primarily for demonstration purposes and the dissemination of good seed, is already on a paying basis, a very unusual state of affairs in a school farm. The buildings and equipment of both farm and school are adequate and excellent throughout. Of live stock there are twenty-nine grade dairy cattle, five horses, eight pure bred sows presented by farmers in return for first choice of one pig from the respective litters, and a flock of hens.

The community is one of prosperous farms on strong, rolling land. General farming prevails, but there is a well-developed and growing dairy interest, and a considerable market garden area in the neighborhood of the county's one city. The attitude of the farmers is distinctly favorable to the school. A local board of farmers cooperates with the state board of industrial education in control of the institution.

The enrolment is eighty-five, some day pupils and the majority boarding pupils, who find their meals and lodging in the village. Twenty-three girls take the work in domestic economy; sixty-two boys, all of whom are from farms, the work in agriculture. The enrolment includes all students, those in the regular courses and those in the winter courses.

The school year is of thirty-two weeks; fall term nine weeks, winter term fourteen weeks, spring term nine weeks. For students who have completed Grade VIII, or for mature students whose evidences of capacity satisfy the principal, a two-year course for the full year is open. Thirty-two boys and young men and four girls have entered under this arrangement. Students of fifteen years of age may enter a course of two successive winter terms. Thirty students are entered in this group. The short year and the winter course are planned in recognition of the fact that farm boys and girls cannot well be spared from home during the growing season—a fact of considerable importance in a consideration of special agricultural schools.

A principal, who teaches, and five teachers make up the school faculty. The principal holds a diploma from a state normal school and the degree of B.S. in agriculture from the state university. He was farm born and raised, and has had twelve years' experience as superintendent and teacher. His salary is $2,000 for twelve months. The farm is in his charge and the demonstration and extension work that is a feature of the summer's plan.

In addition, for the agricultural subjects, there is employed a young man, now in his fourth year as teacher here, who holds the Bachelor's degree in agriculture from the state university. At the university he took the courses in education offered in the agricultural college. He was born and brought up on a farm. Salary $1,250, for ten months.

For shop work the teacher is a farm born man educated at a technical school, an efficient workman, and teacher of four years' experience. Salary $1,350, for twelve months.

The regular two-year course, which carries high school credit for two years, and the winter course follow:

TWO-YEAR AGRICULTURAL COURSE

JUNIOR YEAR

Fall Term		*Winter Term*		*Spring Term*	
Carpentry	4	Carpentry	4	Carpentry	10
Mechanical Drawing	10	Forge Work	6	Types of Animals	2
Dairying	6	Dairy Bacteriology and Sanitation	4	Agronomy	4
Plant Study	3	Types of Animals	4	Horticulture	5
Farm Arithmetic	5	Agronomy	3	English	5
English	5	Arithmetic	5	Bookkeeping	5
Chemistry	5	Chemistry	5	Music	3
Music	3	English	5		
		Music	3		

SENIOR YEAR

Fall Term		*Winter Term*		*Spring Term*	
Stock Judging	4	Feeding of Animals	5	Farm Management	5
Carpentry	4	Farm Mechanics	5	English	5
Field Crops	4	Forge Work	6	Farm Mechanics	10
Horticulture	3	Carpentry	4	Feeds and Feedings	3
English	5	Soils and Fertilizers	5	Civics	5
Building Plans	5	United States History	5	Veterinary Science	5
United States History	5	English	5	Cooking	4
Music	3	Stock Judging	4	Music	3
		Music	3		

Elementary agriculture for girls, 5

On this course the state university allows four units entrance credit.

WINTER'S COURSE IN AGRICULTURE

First Year		*Second Year*	
Farm Arithmetic	5	Feeds and Feeding	5
Carpentry	6	Soils and Fertilizers	5
Business English	5	English	3
Horticulture	5	Farm Crops	3
Animal Husbandry	5	Farm Mechanics	1–4
Dairy	1–4	Carpentry	4
Forge Work	4	Stock Judging	4

So far as possible the winter classes are held with the regular classes, because of the heavy schedule. The English classes, however, are entirely distinct, the regular course English being in accord with college entrance requirements, the short course adapted to real needs.

There is a weekly special lecture, with lantern slides, or moving pictures for all students.

Classroom work is about evenly divided between lectures and text-book recitation. Laboratory work in soils and crops and dairying is individual, in other courses mostly demonstrations by the teacher.

Shop work and drawing are correlated. Students are drilled in the mechanics of drawing before undertaking working drawings for use in the shop, and building plans. In like fashion woodwork and forge work begin with process demonstration and practice and grow through assigned to optional projects, which are the property of students. In mechanics the work in particular is meritorious. Gas engines are very thoroughly mastered.

Of outdoor work none is required except at planting time. For the rest the farm gives opportunities for observation and practice that are pretty largely foregone. In fact, the farm, which is at its best when the students are away, is looked upon as an incubus rather than an aid educationally. Its chief purpose is demonstration for the farming community.

Of home projects there are none other than the keeping of herd records.

The noteworthy adjustments in this school are the conformity of the academic subjects, so called, in the regular course to approved high school standards, and the requirement of a brief course in cooking for boys, and in agriculture for girls.

A Philanthropic Agricultural School

This is a private philanthropic foundation for Jewish youths of New York and Philadelphia, now in its thirteenth year. For entrance an age of eighteen years, the physique for farm labor, and the approval of the principal, secured by personal conference, are necessary. To successful candidates for entrance all privileges of the school are free, including board and rooms. The capacity of the school, 105, is regularly filled, and a long waiting list makes it possible to dismiss students for any breach of regulations or failure in earnest work.

The aim of the school is to make of Jewish lads American farmers, managers, and foremen. Attainment of skill in the art of agriculture takes precedence over acquirement of technology. Because of the lack of expense to the student and because the boys are from the city, where no special summer demand requires their labor at home, the work continues through the twelve months of the year. The course is three years in length.

The usual date of entrance is March, but the beginning of the spring term is with the beginning of the outdoor work, when the soil is fit. When outdoor work becomes pressing, most of the classroom work is given over for the summer term, which lasts till the crops are harvested and the silo filled in the fall, usually about October first. The demands of farm and season dominate in the arrangement of class work. Six days make a week of work, and for farm work during the slack season alternation of classes outdoors and in the classroom is practised. In the forenoon two classes are in the school rooms and one on the farm, in the afternoon one indoors and the other two out. Class periods are forty-five minutes with no doubling.

From the state the school receives $10,000 a year, from Hebrew societies and gifts about an equal amount, and from the farm an income varying from $9,000 to $14,000. The income from the farm in the last year was more than six times its original cost. All work of production, improvement, construction, and repair is performed by students under the direction of instructors or foremen, who are graduates or selected senior students. No building has been erected or made over, no stock or machinery purchased until the distinct need of it has been felt and the money banked in advance. That is, the farm is a business plant developing out of its own accumulation of capital. On this basis it is a remarkably successful farm and a decided financial asset to the school. Against this, however, may be set the fact that a supply of labor, not paid in wages, exists beyond the needs of a farm of the type. The principal states that if he were to allow even moderate wages for much of the necessarily inefficient labor of 'green' city boys, it is doubtful that the farm would show a profit.

Because of this supply of labor, however, and the sane method of development, the four hundred acres of rolling farmlands are in a state of tilth and cleanly cultivation extraordinarily pleasing. Three hundred and fifty acres are in crops, with a large peach orchard that returned a $2,500 crop in the past year, a clean and thrifty apple orchard, vineyard, and extensive gardens. The stock is represented by seventy head of good grade dairy cattle, twenty-seven horses, twenty to 100 swine, according to the season, and 600 to 3,000 fowl. A large motor truck goes to Philadelphia twice a week in the summer and once a week in winter with produce.

All outdoor work of the boys arises out of the needs of the farm and school plant. Rotation in the tasks by groups is the regular proceeding. The boy must master the elementary skills before he can assume responsibility. In the first year his work is for the most part with hand tools, in the second year he may drive a team, in the third year he has his turn in driving the truck to Philadelphia and marketing the product. Even the more difficult task of marketing 'seconds' is entrusted successfully to the boys, who look upon this opportunity as a reward to be sought. For instance, peaches or apples excellent for cooking or canning, but not of the appearance necessary to command a price in Philadelphia are gathered into loads, and the loads assigned to boys singly or in pairs for peddling out in neighboring towns. Only one requirement is set. The load shall fetch a certain small sum, $3.00 or $5.00 as the case may be. Any margin beyond that belongs to the salesman. Never, says the principal, has one returned disconsolate.

Thus the outside work of the school, except as it deals with excess of labor at a given task, comes to correspond very closely with the work of a real farm. The school is well designated a 'farm school'.

To accord with the farm, the school equipment and buildings are plain and serviceable. Except for dairying, the laboratory equipment is small, and the shop is one designed to meet farm needs rather than class purposes. An excellent greenhouse provides not only work for students in horticulture, but products for sale. In the farm buildings not only the labor, but the plans of students have been used. As an instance of the method of growth: The number of young stock had increased to a point beyond the capacity of the barns. Senior students and instructors met for consultation as to whether the surplus should be sold, or a new stable and lintel erected. Because of ample supply of forage the choice was for a building. The boys were then asked to submit plans for a suitable building. From that plan held in conference to be most suitable the building has been erected.

So far does the atmosphere of the farm predominate that classes meet dressed in overalls. But, with that atmosphere there is a disappointing drop from the practicality of the outdoors. Text-book recitation closely competes with the formal lecture. Farm problems no longer control, but 'the nature of the subject'. Laboratory work is at a discount. Farm work and class work run on parallel tracks, and on different schedule, so that the danger of collision is averted as successfully as in many schools where the farm does not loom large.

The course of study as outlined below shows adjustment to season, but little to local demand. On the other hand, the success of the farm has been in no small part due to heed to that demand, from the market and soils angles of selection. The plan for recurrent treatment and the use of physics and chemistry as subjects preparatory to the study of productive agriculture are fairly obvious features of the curriculum, that are not usual:

I

Spring Term		*Summer Term*		*Fall Term*		*Winter Term*	
General Agriculture	5	General Agriculture	3	Farm Arithmetic	3	Algebra	3
Arithmetic	5	Farm Arithmetic	2	Physics	3	Physics	3
English	4	Physics	3	Business Arithmetic	3	Business Arithmetic	3
Botany	3	Farm Work (42 hours)		Botany	3	Botany	3
Entomology	2			Entomology	2	Entomology	2
Physiology and Hygiene	1			General Agriculture	3	General Agriculture	3
Elementary Physics	5			Physiology and Hygiene	1	Poultry	3
Practical Demonstration	2			Farm Work (32 hours)		Farm Work (31 hours)	
Farm Work (31 hours)							

II

Spring Term		*Summer Term*		*Fall Term*		*Winter Term*	
Geometry	3	Surveying	3	Dairy	3	Poultry and Breed Culture	3
Chemistry	3	Chemistry	3	Agricultural Chemistry	3	Organic Chemistry	3
Vegetable Garden	3	Farm Work (48 hours)		Vegetable Garden	3	Nursery	3
Entomology	2			Agricultural Physics	3	Agricultural Physics	3
Fungous Dis.	2			Plant Path.	3	Fertilizers	3
Dairy	3			Entomology	1	Plant Path.	4
Farm Work (31 hours)				Pomology	3	Farm Work (31 hours)	
				Farm Work (31 hours)			

III

Spring Term		*Summer Term*		*Fall Term*		*Winter Term*	
Agricultural Geology	3	Veterinary Lectures	2	Greenhouse Construction	3	Greenhouse Management	3
Forestry	3	Breeding	3	Floriculture	3	Rural Bacteriology	3
An. Physiology	3	Judging	1	Vert. Zoology	3	Cereals	3
Agricultural	3	Farm Work (42 hours)		Cereals	3	Agricultural Literature	3
Chemistry	3			Agricultural Literature	3	Farm Management and Civics	3
Feeding	3			Agricultural Chemistry	3	Farm Work (31 hours)	
Fertilizers	3			Domestic Plants and Animals	3		
Zoology				Farm Management and Civics	3		
Farm Management and Civics	2 3			Farm Work (31 hours)			
Farm Work (31 hours)							

The faculty has nine members, including a practicing veterinarian on part time, and a woman teacher from a high school who comes in for the English. Entomology, botany, and zoology are taught by a woman graduate of a state normal school and student of the university.

In charge of the greenhouse and horticultural work is a practical gardener of many years' experience, trained in the Kew gardens of London. He receives $1,000 and a home. The classroom work is carried on by a recent graduate from an agricultural college. Another is in charge of the poultry classes.

In dairying, an experienced farmer, graduate of an agricultural college, is employed at $1,600.

The teacher in charge of all agricultural work shows an unusual training. He is a Master of Science in agriculture, farm born and experienced, for twelve years professor at an agricultural college, and connected for several years with a state experiment station and the United States biological survey. His salary is $2,000.

The principal, who receives $2,500 and his home, was for many years president of a state college. He studied agriculture, after growing up on a farm, at an agricultural college, as a graduate student at Halle, and took his Ph.D. from Göttingen. His teaching work is in the chemistry of agriculture.

Preparatory School

The single school listed in group C is an endowed preparatory school, founded under religious auspices, but non-sectarian in its enrolment, in which most of the states of the Union and numerous foreign countries are represented. The school year is divided into three terms of fifteen weeks each, and entrance may be made in any term. For entrance an age of sixteen years, an attested good character, and ability to do the work, constitute the requirements. The length of the course is twelve terms, divided into six forms. But the first three terms of work are of an elementary nature. With form 2-b begins work of high school grade, and the courses in agriculture. To this form entrance may be made by passing an examination not far from equivalent to eighth-grade completion, plus a fairly accurate acquaintance with the contents of the English Bible. Many students enter at this stage in the course. In the six forms are 700 boys, all of whom are boarders.

Agriculture is an elective subject by courses. Of these courses there are four: General Agriculture, A, B, C, as shown in the outlines, Horticulture A, B, C, Animal Husbandry, A, B, C, Dairy A, B, each letter representing a term's work of five fifty-minute periods per week. In the outline, the sequence of agricultural courses is such as would be taken by a student enrolled for all agricultural work.

Very few students are taking all the courses, and election results in groups heterogeneous with respect to age and preparation. The agricultural sequence begins with General Agriculture A in the fall term. The academic subjects are those recommended to students by the head of the department of agriculture. These marked with a cross are required. In the agricultural courses, in this, the twelfth year since their installation, are enrolled fifty-six mature students, averaging in age nineteen years, ten months, forty-two of whom are from farms.

The aim of the agricultural work is threefold:

1. To dignify manual labor.
2. To qualify practical farmers and superintendents.
3. To prepare for agricultural and other colleges which give credit for two units of the work:

SECOND FORM B			SECOND FORM B		
Elective 10	† Bible	2	Elective 10	Science	5
	† English	3		United States History	3
	† Algebra	5		Elementary Agriculture A	5
	‡ Modern Languages or Latin	5			

THIRD FORM A			THIRD FORM B		
Elective 10	† Bible	2	Elective 10	† Bible	2
	† English	3		† English	3
	† Algebra	3		† Geometry	5
	† Modern Languages or Latin	5		‡ Latin	5
	Science	5		† German	5
	Civics	3		Mathematics	5
	Farm Crops B	5		Science	5
				History	3
				Farm Management C	5

† Required.
‡ Specially Recommended.

FOURTH FORM A			FOURTH FORM B		
Elective 10	† Bible	2	Elective 15	† Bible	2
	† English	3		† English	3
	† Geometry	5		‡ Latin	5
	‡ Latin	5		‡ German	5
	Science	5		Mathematics	5
	‡ German	5		Science	5
	History	3		History	3
	Horticulture A (Vegetable Gardening)	5		Horticulture B (Fruit Growing)	5

FIFTH FORM A			FIFTH FORM B		
Elective 10	† Bible	2	Elective 10	† Bible	2
	† English	3		† English	3
	† Solid Geometry	5		† Advanced Algebra	5
	‡ German	5		‡ German	5
	Mathematics	5		Mathematics	5
	Science	5		‡ Physics	5
	History	3		History	3
	Horticulture C (Ornamental Gardening and Forestry)	5		Animal Husbandry A	5

SIXTH FORM A			SIXTH FORM B		
Elective 15	† Bible	2	Elective 15	† Bible	2
	† English	4		† English	4
	‡ German	5		‡ German	5
	‡ Mathematics	5		‡ Mathematics	5
	‡ Physics	5		‡ Physics	5
	History	3		History	3
	Ancient History B	5		Ancient History C	5
	Dairy A	5		Dairy B	6

† Required.
‡ Specially Recommended.

The faculty of thirty-six is divided into ten departments, one of which is agriculture, with two men, the superintendent of the farm and head, and a teacher who gives all his time to classroom work.

The farm superintendent and head of the department holds his B.S. in agriculture, and has had six years of teaching experience. He was born and raised on a farm and has had sixteen years' experience as farm superintendent. Salary $2,100 and house.

The classroom teacher holds no degree, but has had four years' agricultural college work and six years' teaching experience. He was born and raised on a farm and thinks and talks in the farmer's language. Salary $1,400 and house.

The head is a believer in the efficacy of 'clear exposition by lecture', but text-book recitation is nearly as frequent in the classroom. This follows the text faithfully, but is lighted up somewhat by discussion in terms of real experience.

Laboratory work, aside from that in the dairy room, plays but a small part. Standard experiments are occasionally set up and demonstrated by the teacher. Of shop work there is, as in the preceding school, only such as may come in assignment to repairs.

Observation and judging, and some practice in pruning and spraying are carried on under the teachers, but the outside work, of which two hours per day is required, including one full day in the week, is for the most part assigned tasks to groups selected in rotation by the head, and under foremen. It thus has no constant or necessary relation to the work of the classroom.

The school plant is a large and excellent one, with the exception of the agricultural laboratory. The school farm is of 1,000 acres, of which 400 acres are in crops, fifty in silage corn, with an apple orchard and large garden. Of stock, there are 175 head of pure bred Holstein cattle, much in evidence at fairs, thirty horses, 100 hogs and a large flock of hens to supply the 700 students and faculty with eggs and poultry 'in commons'. Farm buildings and machinery are excellent. In the library are 118 volumes of agricultural reference and a large file of bulletins and papers.

The school is in the open country in a fairly prosperous dairy region. Soil and market conditions have determined the emphasis on dairying, but the so-called scientific order of presentation dominates in selection and presentation of agriculture. No distinction is made between farm and city boys.

Knowledge of subject matter both in the technological and practical aspects, is marked in the teachers, but the lack of correlation between indoor and outdoor work in circumstances, seemingly peculiarly favorable, is as distinct as in an agricultural college.

AIMS

In every school visited when the principal could be consulted he was asked for a statement of the aims in agricultural courses. In all the schools of Group A the instructor in agriculture was asked for a statement of his aims. In the school of Group B the aims expressed are rather those of the school as a whole than of the courses in agriculture, but, as the work in those schools is dominantly agricultural, the statements are classified with those from the other schools in Table 4.

This classification cannot be considered rigid or exemplary. There is too much vagueness in the expression used, too little uniformity in meaning of the captions, and that even to a considerable extent in schools for which the state department has made a definition of aim. Thus, most of the instructors in the state-aided departments reported the aim of their work as 'vocational', but only five considered their work an entire preparation for productive farming. One principal of a high school, carrying only courses in domestic science and agriculture, and labeled by the state 'Vocational School', stated that his aim was to turn as many students as possible into the state normal schools, and to send his most promising students to college. Since the state college and the normal schools were willing to accept the preparation from his school, and the state paid a large share toward maintaining a school that without aid could not be maintained by the community, he felt that the so-called vocational work should be regarded as preparatory. In general, the meaning of 'vocational' seems to be understood as preparation for beginning intelligently the productive work of the farm. All statements conveying that meaning and those containing the word vocational have been listed in the first column.

The word 'practical' was very frequently used. In general, it seems to denote 'usable knowledge', something that 'they can do with', something that can 'be used on the farm'. But as to what knowledge can be used in practice on the farm, there is difference of opinion. One instructor said of an excursion for the study of geologic origins of soils, "If that isn't good practical agriculture, I don't know what is."

The broader objective of preparation for that mode of life which is called agriculture was stated only five times. That agricultural education implies a preparation for country living is recognized in the stated aims of the Congressional district schools. It does not follow, however, that schools for which an expressed aim, less broad, is given are doing less to the attainment of that end.

Prevocational is a troublesome word, but was used six times to express the aim of high school agriculture. To be sure, one principal expressed the aim as "prevocational, if any," but the others were more certain of their object, "To gain a sympathy with and understanding of farm life," another put it. "Units of science that may give an opportunity for intelligent selection of a calling," conveys the idea of predilection that seems a part of the idea expressed by others, and in particular of one experienced head of agricultural education in a great state. His words were to this effect: "There are three aspects to the aim of agriculture in our high schools: the prevocational, the vocational, and the liberalizing. It should give opportunity for sympathetic and intelligent election of an occupation; preparation for the productive activities of that occupation, and insight as to the implications of a great social activity."

Only four times was the object of an enlarged intellectual and emotional appreciation, the liberal or cultural aim, expressed, though 'culture' was in no case defined. On the other hand, only twice was doubt expressed as to the 'cultural' value of agriculture. One principal said: "We strive for culture. We do not advertise our agriculture. It is for the local farmers' sons." Another said he much regretted that he had not been in office at the time of the establishment of the work in agriculture, so that he might have forestalled its introduction. He intended to give his best efforts to a reduction of the time devoted to it, not that it was not useful, but because 'culture' was more worth while.

The college preparatory aim is evident in more cases than those in which it was actually stated. In no case was it given as the primary aim of agricultural work. In the study of the curriculum the marked influence of the preparatory idea will come out.

Other aims expressed were preparation for 'citizenship', or the 'active duties of a citizen', 'to reach the boy', 'to reach the parent', 'to dignify manual labor'. And two principals rejoiced in the introduction of vocational agriculture as an 'entering wedge to the overthrow of a tyrannous system' of uniform state examinations. This they regarded as, if not an end, at least a justification for agriculture in the curriculum.

EQUIPMENT

Table 5 shows the possession of land by the various schools, either as plots owned, hired or borrowed, or as a school farm. It will be noted that nineteen of the high school group have the use of some land, and seven own farms. Of the other groups, all have farms. The amount of land in plots owned, hired or borrowed for agricultural purposes varies from a small hen-yard in a town high school to twenty acres in a city high school. Other town or district high schools report one-fourth acre, one-half acre (two schools), one acre (two schools), one and one-half acres, two and one-half acres, four acres, nine acres and ten acres respectively. Other city schools report one-fourth acre, one-third acre, one-half acre, eleven acres respectively; county high schools, two acres and seven acres; academy, three acres. Three academies have farms of thirty-five acres, and 150 acres (two); Congressional district schools, one sixty acres, two 250 acres, one 312 acres; county agricultural schools, one eighteen acres, one 126 acres, one 136 acres, one 130 acres, of which ninety acres are still under lease to a seedsman and not available for school uses. Of the state schools, one has a farm of 200 acres, one of 2,000 acres; the other is given the use of the university farm and equipment of the state. The philanthropic schools have farms, one of ninety-three acres, one of 400 acres, one of 600 acres. The preparatory school has a farm of 1,000 acres. The uses to which plots and farms are put will be discussed briefly later.

Table 5–A shows the possession of neat stock, cattle, swine, or sheep, horses, or mules, poultry, and farm machinery. The parallel with respect to possession of farms is fairly exact in the schools of Groups B and C. The state schools are adequately, two of them even magnificently, equipped with stock and machinery. For instance, one state school owns 250 head of pure bred cattle, forty horses, 100 pure bred swine, and all the necessary modern machinery of a 2,000-acre farm. Three of the country agricultural schools have an excellent equipment of machinery and satisfactory live stock. The other is run on a home project basis, and has a relatively small equipment. The philanthropic schools are equipped, two of them on a business basis only, but have the necessary machinery and stock for good farming. The other is on the home project basis and has largely foregone the possession of live stock. The preparatory school has 175 head of pure bred dairy cattle, thirty horses, and an excellent business equipment.

In Group A, the possession of live stock is at a minimum. Poultry is the most frequent item. The Congressional district schools, though possessed of large farms, are conspicuously understocked and under-equipped. Fifteen to twenty cattle of inferior quality, on three farms, and one bull on the other represent the cattle. Some good mules and pure bred hogs were found on these farms, and one flock of pure bred poultry. Only one had the farm buildings and machinery that may be found on any moderately prosperous northern farm of the size. In respect to use of land, these schools make a much better showing than the equipment would indicate. They are fighting an uphill fight against poverty.

Of the farms owned by academies, only one really deserves the name of farm. The others are tracts of land on which the school stands. That one farm has a small well-housed herd of good dairy cattle, a pair of good horses, nondescript hogs and poultry, wallowing in filth, and a very modest equipment of machinery and buildings. Good poultry is kept by another academy; the third has a handful of grade dairy cattle. In the town and county high schools represented, the poultry consists of pure bred flocks. The machinery varies from a single gas engine to a room stocked with modern horse implements donated by warehouse concerns.

Two city high schools and one town high school have small greenhouses. One is "a place to keep ornamental plants when they are not needed for parties," another is used for classes in botany, and the third is a place in which students in horticulture grow tomatoes, peppers, cabbage and other plants for sale. Seven of the schools in Group B have greenhouses that are in regular use by agricultural students. Six schools of the high school group have placed the agricultural classrooms and equipment in separate buildings. In one case, this building was constructed for the purpose; in another, it is a rented store, well fitted up; in the others, school buildings which have been abandoned for other purposes. In eight cases, all agricultural work is conducted in the basement. In one county school and two city schools the instructor in agriculture was allowed no classroom of his own.

Twenty-seven of the thirty-nine schools in the high school group have some shop equipment. Twelve are equipped for woodwork only, ten have forge as well as carpentry fittings, and five have small farm shops with a bench and tools, and in three cases, a forge. The equipment varies from rough benches built by the boys with their own tools to the elaborate manual training rooms of the city high school, with electric power and individual benches. The equipment listed in the first two columns is classed as scanty in the case of five town or district high schools, three of them state-aided, in one county high school, one Congressional district school, and one academy, state-aided; as adequate, in the case of one city high school, state-aided, five town or district high schools, state-aided, one county high school, one academy, and one Congressional district school; as excellent, in the case of three city high schools, two state-aided, and one Congressional district school.

In Group B, forge and woodwork equipment, with power, is found in nine schools. This equipment is rated adequate in the case of one county agricultural school and one philanthropic school; as excellent, in the case of the others.

Table 5 shows the possession and rating of laboratory equipment in all schools. Of the five schools listed as having no equipment, four had equipment of some sort for chemistry or physics. The two philanthropic schools rated as having scanty equipment were adequately equipped for dairy work, and the preparatory school excellently so equipped. Poverty in respect to laboratory material is marked in the schools of Group A, twenty-two of the thirty-nine having only scanty equipment, or none at all. The case is particularly marked with the Congressional district schools, only one of which is possessed of equipment, and that very scanty. On the other hand, as in the matter of stock and machinery, and shop equipment, the schools of the special type, Group B, stand in rather marked contrast.

Tables 6–A and 6–B show the possession of bulletins and volumes for reference in agricultural subjects. The number of bulletins owned by the school is not recorded. At the start of the investigation a count was made in several schools, but the time required was too great. With bulletins coming in at frequent intervals, it was not astonishing that no instructor knew the exact number of bulletins in his possession. In general, the classified lists were larger than the scattered. In the schools of Group B, the bulletins may be numbered even in thousands. Two schools, both of which possessed classified lists, required the students to send to the state agricultural experiment station and to Washington for such bulletins as must be studied. Opinion prevailed that books are more satisfactory than bulletins for reference use. Every instructor of whom the question was asked affirmed that opinion, but the question was not asked of more than thirty instructors. In the matter of reference books, three schools, one city, one town, and one Congressional district school, reported dependence on public libraries, but did not know how many volumes on agricultural subjects were available in them. In Tables 6, 6–A, and 6–B the same general differences between Groups A and B as in the other equipment tables are observable. In the matter of equipment, as might be expected, the special schools of agriculture have advantage of schools of the high school type.

In forty-three schools, lists of agricultural reference books were secured, and in all schools lists of the text-books actually in use in agricultural subjects so recognized. Of different titles there appear in the lists 628. Counting only one volume to each school, either as text or reference, the total comes to 2,226, of which 1,978 are reference books, 248 texts regularly in use. Variation in the number of reference volumes is shown in Table 6–B. In Table 7 are shown the number of titles under the topic treated and the number in each such group now in regular use as text-books. The distribution may suggest confirmation of the statement made previously concerning the productive aspect of the mass of country life publications. A list of titles appearing in five or more schools as reference or texts is given also. (Table 8.)

TEACHERS

Among schools of the high school type, in Group A, teachers of agricultural subjects are ordinarily called upon to teach subjects other than those recognized as belonging to the field of productive agriculture. In twenty-six of the thirty-nine schools the teacher of agriculture is so employed. The granting of special aid for agriculture to the high school usually entails the requirement that the teacher shall give all his time to agriculture. In Maine and Wisconsin, the requirement is not made. In Minnesota, the amount of state aid is diminished in the ratio that time given by the agriculture man to other subjects bears to the total of his teaching hours. But in other states the requirement is not strictly enforced. Of the twenty-six schools in which teachers give part time to other classes, ten are receiving grants of state aid particularly for the maintenance of agricultural courses. It is noteworthy, however, that for the thirteen schools in which all teaching time by those men is given to agriculture, special state aid is granted. Two or three added subjects are common.

In a school not receiving state aid for the special subject it is not ordinarily possible to employ a teacher whose whole time shall be given to the teaching of the one subject. In schools receiving special aid the newness of the installation of the work may mean that the teacher is not yet fully employed by the tasks of his own department, and that he is, accordingly, drafted into other work.

Commonly, the teacher of agriculture, even in the high school group, gives a part of his time to extension work outside of school hours. In twenty of the thirty-nine schools of Group A this fact appears. He may also have charge of garden work in the grades, or even of the teaching of agriculture in the grades, as is shown in the same table.

In only five schools of the first group did the teacher of agriculture serve as principal also. Most commonly he acted, if called upon for other subjects, as teacher of the sciences, botany, physics, zoology, chemistry, or general science, but the range of subjects taught by high school teachers of agriculture is wide, as may be seen in Table 8.

In the fifty schools visited, the regular teachers of agricultural subjects were men, in the schools of Groups B and C strictly departmental teachers; in the schools of the high school group more widely distributed as to duties. One county high school, and two town high schools state-aided, employed two men for the agricultural work, but the rule in the high school is the employment of a single teacher. It is not to be inferred, however, from the results of the study that teachers of agriculture in the high school are always men. In Table 1–D, dealing with the distribution of salaries among teachers of agriculture in the high schools of Iowa, where no state aid is given, it is shown that fifty-eight out of the 406 teachers in the list are women. In the half-unit and one-unit courses common in middle western rural high schools, where the text-book dominates in company with the state manual, there is no reason why an intelligent woman, even from the city, should not conduct the course in approved fashion as well as the male teacher of science, the principal, or the superintendent. But in the more promising schools considered in the study, the tendency to employment of specially trained men is marked.

In Tables 9 and 9–A the training of seventy-nine teachers of agriculture, forty-two in Group A and thirty-seven in Groups B and C is shown, together with the visitor's rating of the teaching ability demonstrated at the time of visit. This rating can hardly be regarded as conclusive in any particular. It is, of course, impressionistic, and subject to the variations in mood, and the personal attitude of the visitor. Moreover, it is based upon but a single visit. Yet it was made with certain criteria in mind, particularly, knowledge of the subject displayed, specific applications in discussion, concrete and thought-provoking questioning, evaluation and emphasis, evident motive and initiative in the class. In certain subjects, particularly in the field of agricultural engineering, the visitor is not qualified to pass judgment upon the knowledge shown by teacher or class, or upon the evaluation and emphasis given to different phases of the work. Accordingly, teachers in these subjects have for the most part not been rated. If any suggestion is to be derived from this rating, it is that the somewhat superior training of teachers in the schools of Groups B and C makes itself evident in the classroom.

As factors in training are listed, birth and upbringing, duration of farm experience of city-born teachers, years of teaching experience, length of pedagogical training, preparation in subject including, for shop teachers, the 'school' of trade practice.

It will be noted that of the seventy-nine teachers in the table, sixty-five were born and brought up on farms. In the high schools eleven of the forty-two were city or town bred, of those in the special schools only three of thirty-five; more than twenty-five per cent. in the one group, less than ten per cent. in the other. Most of these city-born teachers have had some farm experience, only three reporting none outside of that connected with their agricultural college training. An experience, however, only in summer work is one by no means equivalent to that which takes in the winter aspects of farm living also. The teacher whose record shows an experience of six summers had undertaken consistently to make up what he felt was a defect in his training. During the eight summers he has had free since his graduation from college, he has spent six as a hired laborer, each on a different farm in a different locality. In the other two, he made trips to the middle west and Pacific slope and traveled afoot through farming sections. Among the farm-born teachers are several who have been for from one to twenty years, managers or working owners of farms.

In teaching experience, many of the men are deficient. Twelve are in their first year, forty-five have taught three years or less, of whom thirty-three are in the high schools. In length of experience the advantage lies with the teachers of the special agricultural schools, twenty-two of the thirty-five having had from four to twenty-two years of teaching.

Twenty-six of the seventy-nine teachers have had some measure of pedagogical training, varying from a single summer's course in education, to the completion, in one case, of the requirements for the Master's degree in educational subjects. With respect to training in the teaching process, the high school group makes the greater showing. Eighteen teachers have taken courses in education, twelve of them during the last year of their college course.

Agricultural college training is the rule, thirty-seven of the forty-two teachers in Group A reporting such training; in Groups B and C, thirty-two of the thirty-nine, seven being shop men of trade or technical school training. The degree of B.S., given by an agricultural college is by far the most common, appearing fifty-three times, thirty-four in the high schools, nineteen in the other groups. Master of Science in agriculture appears once in the high school group, six times among the agricultural schools. The degree of Ph.D. is held by directors of three of the agricultural schools.

The typical teacher in the high school group holds his B.S. from an agricultural college, was born and raised on a farm, has taught school less than two years, and has had no pedagogical training. The same may be said of the teacher in the agricultural school except that his experience in teaching is five years and he is more likely to have achieved a higher degree. The generalization must not be carried over into the whole field, however, though the difference would become more marked.

Under the home project scheme common in state-aided high schools, and in the agricultural schools where extension work is usually a duty of the teacher, he is likely to be hired for twelve months, rather than for the academic year; fifty-nine of the seventy-nine teachers are so hired. In the unaided high school such tenure is probably the rare exception.

SALARIES

Table 10 shows the range of salaries among 406 teachers of agriculture in the high schools of Iowa, where no special state aid is given. In the data available there was no means of determining whether or not the teacher of agriculture was also the principal or superintendent. That he is, is doubtless the case frequently, and tends to skew the distribution upward. In Iowa, it would appear that it is more profitable to be a man than a woman, more profitable to have some agricultural college training than to be without it. For all teachers of agriculture $900 is the favorite allotment of salary; for women about $650, for men from agricultural colleges about $1,000, for women from agricultural colleges about $700.

The effect of state aid upon salaries may be noted on page 26-A where are shown salaries of teachers in sixty-one state-aided schools of New York in 1915. The median salary is there $1,200. All these men have an agricultural college training or what is approved by the state department of education as equivalent. A like effect appears in the distribution of teachers' salaries in 136 state-aided high schools of Minnesota in 1913–1914. The median there is $1,250, and salaries have been 'somewhat increased' since then. The same statement in regard to training holds as with the New York teachers. Classification by colleges in which state-aided high school teachers of agriculture were trained, as reported for 1915, is indicative of state requirements under grant of state aid. See page 28–A.

Salaries reported from four county agricultural schools of Wisconsin for 1913–1914, including salaries of directors, show a range from $850 to $3,000, with a median salary of $1,500. Because of the form in which figures for all these schools were given, it was possible to use only the four; but the median is not far from correct for all schools.

For the fifty schools of the study the distribution of salaries for seventy-seven teachers is shown in Table 12. In the high schools the range is from $540 for the teacher of science and agriculture in a Congressional district school to $2,100 for the teacher of agriculture in a state-aided town high school. The median salary is $1,200. In the agricultural schools the range is from $840 for the instructor in dairying in a philanthropic school to $5,000 to the director of a school of the same group. All salaries above $2,000 are paid to directors. The median salary for this group, including salaries of directors or principals, is $1,600; for directors, $2,700; for other teachers, $1,500. The figures in the table include estimates for board or rent when such make a part of the compensation. Teachers in the special schools again appear somewhat better paid than those in the high schools.

ENROLMENT

A tabulation of the enrolment in agriculture in 462 high schools of Iowa shows a median enrolment of twenty pupils. The same median appears in the enrolment for forty-four state-aided high schools of New York. For the schools of the study the range in Group A is from ten to 143 pupils, the larger numbers appearing in the Congressional district and county schools where every boy takes agriculture, and, in one case, every girl, also. The median stands at twenty-one. In Group B the range is from forty-one in a philanthropic school to 597 in a state school. The median is 105.

As might be expected in the special schools where attendance implies the intention to study agriculture, the enrolment is larger than in the high schools where various courses are open. In only one school of Group B can a student gain his diploma without taking agriculture. That school shows the smallest enrolment in that course of any in the group.

A trend to the selection of farm lads is probably indicated in Table 13. This is, no doubt, in part due to location, though even one city high school shows over ninety per cent. farm boys in agriculture. On the other hand, it may be noted that three city high schools are dealing with city boys, whereas the Congressional district schools in the open country in states where cities are few, deal with country lads. The two philanthropic schools in column 1 were founded for city boys, the one county agricultural school showing a majority of city boys is located in a suburb of a large city. The median proportion of country lads is about sixty-five per cent. in high schools, in the special schools about seventy-five per cent. The preparatory school with a cosmopolitan enrolment of 700 boys, shows a distinct selection of the country lads.

An important pedagogical problem is indicated in the fact that more than half the schools are dealing with groups of boys of widely different life experience before entering school. In what respects adjustments have been made to meet this problem will be considered later.

AGE OF PUPILS

The impression gained by the visitor was that the Congressional district schools in Group A and the agricultural schools were dealing with more mature students than the run of high schools. The ages given in Table 14 seem to confirm the impression. The figures, however, are not strictly accurate, more than half the returns being obvious estimates by teacher or principal. Three factors enter to account for the greater maturity of students in these schools: (1) The higher age requirements for entrance; (2) the fact that all but two of the schools are 'away-from-home schools', to which immature students are less likely to be entrusted; (3) the greater diversification in agricultural work that attracts students who have been at work for some time already. Men of twenty-five or more were not infrequently in the classes of such schools.

Upon this greater maturity several principals have based the use of the lecture method and the close approximation to college procedure that distinguish the schools as a group from the high schools.

ENTRANCE REQUIREMENTS

Entrance requirements to courses or departments in agriculture are commonly identical with those for other courses offered in the school. However, in the state-aided vocational departments of high schools in Indiana, Pennsylvania, New York, and Massachusetts, provision is made for the entrance to work in agriculture of those who do not seek a high school diploma on an age basis of fourteen years. Fifteen schools in Group A, all of them state-aided, have this provision. Eleven of them show the admission of students with the approval of the principal on the age basis. The number so admitted is, however, small. In the Congressional district schools students are in practice admitted who have not met the established requirements. With the approval of the principal, students of high school age, considered capable of carrying the school work have been admitted on trial.

In one country agricultural school and the three philanthropic schools, students are admitted on the age basis alone. In the philanthropic schools, age is the basic requirement. In one, 'fourteen years and the desire to take the course', is all that is necessary; in the others, 'eighteen years and the physique for farm labor, with ability to read and write'. In the latter two, selection of candidates is made by the principal through personal interview. In the state schools, two require an age of sixteen years in addition to the completion of Grade VIII or a preparation approved as equivalent, one, an age of seventeen years and not less than six months' farm experience besides the completion of Grade VIII. In three county agricultural schools, fourteen years and the approved equivalent of Grade VIII completion is necessary.

In general, the completion of elementary school work in the particular state wherein the school is located determines entrance, the fact of completion being determined sometimes by certificate, sometimes by examination. Thus, thirty-one schools of the high school group base admission to agricultural courses on Grade VIII completion, and seven of the schools of Group B do the same. Requirements for School C are already given. Grade VIII completion probably represents admission standards for the great majority of schools in the country, Grade VII being standard in the South.

UNITS OF AGRICULTURE OFFERED IN HIGH SCHOOLS

In schools of the high school type, offerings in agriculture have been reduced approximately to the common high school unit, for exhibition: *i. e.*, five forty-minute periods per week through the academic year, a double period of laboratory or shop work counting as one period for credit. The seven and one-half Regents credits for agriculture in New York thus are one and one-half units per year, or six units for the course. The fifteen periods a week for agriculture in Massachusetts and Pennsylvania count for eight units in the four-year course. Table 15 shows the distribution of schools in Group A under this unit classification.

The distribution of the same schools under state aid for courses in agriculture, state maintenance of school irrespective of agriculture, and without aid from the state, appears in Table 2.

The effect of state aid in increasing the number of units offered is evident from Table 16. The median offering for the sixteen schools in which no special aid is given for agriculture is 4; for schools to which such special aid is granted, 6.

Additional figures suggest a like effect.

In Iowa, 492 high schools, unaided, show the following distribution of units, with a median offering of one unit:

UNITS	SCHOOLS
½	243
¾	3
1	212
1½	6
2	19
2½	1
3	3
4	5

In Illinois, of seventy schools, unaided, carrying agriculture, ten carry two units or more.

In Vermont, slightly aided schools, nine offer four units, four offer two units.

In New Hampshire, unaided schools, one offers six units, twenty-two offer four units, two offer two units.

In Michigan, unaided schools, fifty have installed a four-unit course.

In Minnesota, 176 state-aided schools have met the state requirements calling for a four-unit course, though the actual offering is sometimes less.

In New York, sixty-one state-aided high schools are offering six units, and three state-aided intermediate schools devote the same time to a like course.

In Pennsylvania, twenty-three, and in Massachusetts eleven high schools are receiving aid for an eight-unit course.

For the agricultural schools with their more specialized courses, greater diversification of agricultural subject matter, less uniform length of year, and of courses, no attempt at evaluation in terms of the high school unit has been made. The total offerings, however, are normally higher than in the high schools.

The greater uniformity in length of year and course in the high school group is shown in Table 17. Apparently, state aid has no effect to lengthen the course, except as the summer projects be included, as will be noted later. Normal length of year is thirty-six weeks; of course, four years. In the agricultural schools variation in length of course and year is marked, the range being from one year to four, from twenty-four weeks to twelve months.

The presence of boarding pupils, though it presents a problem in the matter of maintenance of projects, has no effect on length of course. It is noteworthy, however, that the only schools to maintain a school year of twelve months are boarding schools.

THE COURSE OF STUDY—THE ACADEMIC SUBJECTS

In the following brief discussion of those subjects in the curriculum which are not recognized, usually, as agricultural, there are included in the compilation several which are normally elective. In such cases the basis for inclusion is the fact that they are those taken by a majority of students in agriculture or are regularly recommended for such students by principals or directors. In three schools, where long lists of electives were provided, the data were not sufficiently clear to be included, and only those subjects that were required of students in agriculture enter in the tables.

English

Of all subjects in the course of study, English is the most uniformly required. Every school of the fifty offered English; only one, a philanthropic school, failed to require it. In the units offered as well as in the content, the schools of Group B show some diversity; in the other schools uniformity is notable.

Thirty-six of the thirty-nine high schools offer four years of English, the other three offer three years. Thirty schools offer four full units; four, three units; four, two and one-half units; one, two units. Two of the four-unit schools make the subject elective in the last year. In all others the full offering is required.

In content, the subject is usually college entrance English, consisting of composition under the heads of Narration, Description, Exposition and Argument, and the shredding of literary masterpieces chosen from the list in college entrance requirements or the state manual. In the first year, or first two years, part of the time may be given to English grammar, as is notably the case in the Southern county and Congressional district schools, and the one junior high school.

No case of differentiation of English to meet special needs of agricultural students is recorded among the high schools unless the use of the Silver Burdette *Farm Readers* in the Congressional district schools be noted as such.

The smaller number of units noted for the agricultural schools is due mainly to the shorter courses and shorter years in those schools.

High school 'standards' regulate the content for the most part. Only for short-course students in two country agricultural schools, and for all students in the philanthropic schools is special English given. The course then becomes 'Business English' or 'Vocational English' and deals with the writing of business letters, the summarizing of texts, articles or bulletins on agricultural subjects, and the reading of country life literature, or of masterpieces for the interest that is in them. In one case the course is designated 'Conference English', and is supposed to help students in the organization of their notes and readings in other classes. Mr. Perry of the Milwaukee County Agricultural School has organized his work in this wise:

Conversation and Reading, Correspondence and Composition, Current Literature of Farm and Home, Organization and Debating, a term being given to each topic.

Reading

Reading of the regular grade selection is given in the first two years of the junior high school, and in the first year of one state-aided high school of New York.

Spelling and Writing

Penmanship is required for two periods a week in the third year of one county high school, and for the first year of one Congressional district school, after which the same time is given to spelling in the following two years. Two other such schools give one period to writing and spelling through two and three years respectively. One New York school requires a period during the first year, and in the junior high school it continues through the full course.

Mathematics—Arithmetic

Arithmetic, usually a review of processes, is offered in thirteen schools of Group A and seven of Group B. In only four cases in each group is it differentiated to meet the needs of agricultural students, when it becomes 'Farm Arithmetic', usually an attempt to teach agriculture through arithmetic, rather than to teach arithmetic through the farm experiences of the pupil. The work, however, may be more or less concrete and applicable. The notable exception is the junior high school, where the work is carried on through the first year without an assigned text-book. In project schools a good deal of concrete arithmetic enters into the work even though no definite course be given. The same can be said of courses in feeding and fertilizers.

Algebra

Algebra to quadratic equations is the rule in the high schools, thirty-six of the thirty-nine requiring the subject, and two offering advanced or review work in the subject. In the county agricultural schools it appears as a factor in receiving high school credit, in the philanthropic school a brief dose is given as tonic to the mind.

Geometry

A like statement is applicable to geometry. In the solid form it is included with Review Mathematics. Strangely enough, an agricultural student in the preparatory school may escape it, though his chances are not great, because of the pressure of college preparation.

Surveying

Surveying is offered in six schools, all of which have farms and carry on actual practice in land measurements. In two cases leveling and the laying of drains in an actual project was reported. A brief treatment of farm surveying is frequent in farm management courses.

History and Civics

United States history and government is as regularly a required subject as algebra or geometry. It is regarded as of peculiar value as a preparation for citizenship. At the same time it would appear that only those boys who survive the first three years of the high school are thought worthy of this particular training. Only six times is the work offered earlier than the fourth year of the course. In the special schools where the course is of two years only, it becomes necessary to give the subject in the second year.

United States history and civics may be given as separate subjects. The United States history thus given appears to be the regular eighth grade history, in schools where Grade VIII is included. Civics as a separate subject appears more frequently in the special schools than in the high school group and is sometimes differentiated from the ordinary study of national, state and local government. In one case it is studied in connection with sanitation, in another designated as 'Community Civics', in another as 'Community and Citizen'. In one philanthropic school for Jewish boys, civics, like English, is elective, and is seldom elected. In another, civics is a part of the farm management.

Ancient, English and general history form a part of the work of agricultural students in many of the high schools, English history being most frequent. State history is required of students in the Georgia Congressional district schools, and in the county high school of Maryland. Industrial history and the history of agriculture appear in only three schools, one county high school and two county agricultural schools. The history of his vocation is considered of less importance to the student than that of the wars of foreign peoples.

Fine Arts and Music

Fine Arts is required of agricultural students for two periods a week in one New England academy, for cultural purposes. Music, which is worthy of serious attention on the part of one who plans a course in preparation for the life of a farmer, is given to students in two county agricultural schools throughout the course, in one case one period a week, in the other, three periods. One county high school teaches a class one period a week for two years, a town high school two periods a week for the first year. The allotment of time to music does not seem extravagant in schools which incorporate in their requirements, algebra and Latin.

Foreign Languages

The tendency to the omission of foreign languages from the course for agricultural students is interesting. Latin is given them regularly in only five schools, French in four, and German in eight. The schools in which French is taught are in New England, all of them in New Hampshire, where the French Canadian population is large. Five of the schools that teach German are in the Middle West, in communities where people of German origin are many. The most notable point in regard to foreign languages is the complete omission in the special schools.

Geography

Geography of the eighth grade is taught in two schools. Commercial geography, in which the correlation with the productive side of agriculture is marked, is taught in one high school. Physical geography and 'Geology of the Farm' go together in one county agricultural school.

Bible

Bible study is required of all students in one New England academy, and in the preparatory school, both of which were founded under church auspices. It is elective in one philanthropic school, but is seldom elected.

Gymnasium

Required gymnastics throughout the course appears in two schools only, one a state school, where actual farm work is at a minimum. Directors in the special schools hold that farm work does away with the necessity for gymnastic exercise. In the high schools athletic games were usual, sometimes with a special coach, and several had well equipped gymnasiums, but physical exercise did not appear as a requirement.

The Sciences

In the study of the production of plants and animals, the principles of science must be applied to control of the race through heredity, of the individual through nurture. There are three views commonly expressed with regard then to the place of the sciences in the course of study for students of agriculture. *One:* That the function of the science courses is to furnish a background in classroom, laboratory, and field experience, for the technology of agriculture. *Two:* That the sciences should furnish an organized structure of correlated natural laws that shall be clothed and filled in through the concrete applications involved in the later or concurrent study of agriculture. *Three:* That the concrete applications of science in agriculture shall be an inductive means to the discovery of the laws of nature, which shall later or concurrently be unified in the study of pure science. In other words, the first two views look upon science as preparatory to agricultural subjects, the third looks upon agriculture as preparatory to science. These views were expressed by some eight or ten principals and state officers, as guiding in the organization of the curriculum. The one set would indicate the placing of science early in the course, the other late. But that such views do actually dominate in the run of courses is doubtful, for if such is the case the maker of the course ordinarily holds both views. That is, in the same course, commonly, botany is given in the first year, chemistry in the last. But that certain sciences are predominantly looked upon as preparatory, may be judged from the following synopsis:

General Science

One course that is undoubtedly given its place for the sake of its 'background' value is General Science, that much quarreled over hodge-podge of biological, chemical, and physical instances. It finds its place, however, in relatively few schools, only seven of the fifty. In every case it is given in the first year, either as a whole or a half unit. As a half unit, it is in one case, preceded by a half unit of elementary agriculture, in the others, it is the first study of science.

Geology

Geology is a rare offering, appearing in but three schools. In one high school it appears only with astronomy as an informational subject; in the county agricultural schools it is regarded as an adjunct to soils work, as it well may be. Some study of geology is regularly made in soils courses.

Botany and Zoology

Botany, usually as a half unit preceding zoology, appears normally in the first year of the course, and rarely in the second or third year. It is a common requirement appearing in twenty-six of the high schools and ten of the eleven schools in the other groups. In three schools of Group A and three of Group B it is differentiated as 'Agricultural Botany' or 'Farm Plant Life', in which case the differentiation consists mainly in the selection of plants of economic value, and a greater emphasis on physiology, than in the usual structural botany of the high school texts. How crops grow is of more importance to the farmer than the structural variations that they exhibit. How they feed is more important than the arrangement of leaves upon the stem or the presence or absence of petals in the flower. These facts seem to be recognized in the adapted courses. But such courses, as the regular ones, are held to be preparatory.

Zoology, in its distribution, is like botany, though a less frequent offering. In only two cases was adaptation definitely made for agricultural students. Its inclusion in two more special schools as 'Zoology of the Farm' brings the total for those schools to three as against nine for botany. But the adapted courses seemed so closely to resemble those in economic entomology that they have been listed under that heading.

Physiology and Hygiene

Physiology and hygiene of the usual text-book type was found in seven schools. It is equally common in the first and second years and varies from a full unit to one-fourth unit or less. The content seemed to be that of the usual eighth-grade text. It was the visitor's privilege to listen to three recitations and one sermon on the evils of drink.

Sanitation

Sanitation as a separate division appears in the Southern schools and in two of the special schools. In three schools boys had constructed concrete septic tanks, though in one such case, no course in sanitation was offered. In the seven cases where it was offered it seemed rather direct in its application to the farm home, even when given as 'Civic Biology'.

Physics and Chemistry

Physics and chemistry are commonly offered as a single unit each in the third and fourth years respectively of the high school course. In the smaller schools the usual practice, where both subjects are offered, is to give them in alternate years, combining the work of junior and senior classes. The sequence shown then, is not strictly representative. It is based upon the printed courses of study, or in the absence of such, upon the actual succession in use at the time of visit. Physics is somewhat more common in the high schools than chemistry, appearing thirty-seven times as against thirty for chemistry. In the special schools, physics appears six times as against nine for chemistry. The preparatory view in the special schools is indicated by the earlier appearance in the course, physics appearing four times in the first year, chemistry six. Differentiation into Agricultural Chemistry or Agricultural Physics is rare in both groups. In some cases the work was said to be given an 'agricultural bias'. It will be noted that both sciences are among those most frequently in charge of the teacher in agriculture. But the 'bias' in classes visited was imperceptible. The usual statement was that 'the entrance requirements forbid differentiation for agricultural students'.

No record of texts used was kept for subjects other than agriculture, but Millikan and Gale's text, with the accompanying manual, or some outline like the "Forty Harvard Experiments" was in common use in physics. In chemistry, the usual inorganic chemistry, based on Newell or a like text, was the rule. In cases where agricultural chemistry was offered, Kahlenberg and Hart or Snyder's "Chemistry of Plant and Animal Life" determined the content. In agricultural physics, King's "Physics of Agriculture" was twice used. In one case the course could hardly be distinguished from that in agricultural engineering. Just what is meant by agricultural chemistry, physics or botany seems not yet to be certain in secondary schools.

One specially ambitious development of chemistry, from a special school, may be of interest. Elementary chemistry, two terms, elementary agricultural chemistry, one term, elementary organic chemistry, one term, advanced agricultural chemistry, two terms. It is noteworthy, however, that laboratory work was at a discount in this course. The instructor expressed a doubt as to the value of such work, except as it were confined to illustrative demonstration in the classroom, in immediate conjunction with classroom recitation or lecture. This man had taught boys in the subject for more than twenty years, and had taken his doctor's degree in the subject from one of the German universities. If laboratory work be as remote from the content of instruction as is usual under the manual and outline method, no doubt there is some virtue in his contention.

BOOK-KEEPING

Book-keeping, usually in the second year, and as a half unit, appears as a separate subject in nine high schools and five special schools. Seven times in the high schools, and in every instance in the special schools, it is adapted more or less closely to farm conditions as 'Farm Accounts' or 'Vocational Accounts'. In schools where farm management is taught it sometimes becomes the major work in that subject. In schools where home projects are a part of the agricultural course an accounting of productive projects is invariable. Thus, in a majority of the schools visited, boys become acquainted through practice with simple accounting.

Agriculture for Girls

Girls study agriculture in seven of the schools, six high schools and one special school. In one county school a first year unit is required of them, in one Congressional district school three and one-half units. A text-book in general agriculture and stereotyped recitation is the type. In three cases, however, the material is selected with a view to the supposed requirements of the home-keeper. Vegetable gardening, home fruits, poultry management, floriculture, and the care of home grounds, make up the work, which is given to a separate class, away from the boys. In one case, only, was there any noticeable amount of practical work. In that case every girl carried a project. The class visited homes, conservatories, fairs, poultry yards and orchards, designed schemes of ornamentation, selected varieties of fruits and vegetables, in fact, gave most of the double class periods in spring and fall to out-door work. Yet the teacher reported the work unsatisfactory, and rejoiced in the prospect of a cooking and sewing class that should remove the girls from his jurisdiction. They did not care for the practical work, he said, and were at their best in text recitation.

Cookery for Boys

An interesting special selection is that of camp cookery or plain cooking for boys, during a term in each of two Wisconsin county agricultural schools. The necessity for preparing his own meals at times is almost inevitable in the life of a farmer. If anywhere selection of subject matter as usable knowledge appears, it is here. A brief course can hardly be out of place in any school that prepares for country living. In another school the boys had formed a cooking club.

Drawing

Drawing as a special division of the work of students in agriculture appears in sixteen schools, ten in Group A, and six in Group B. The arrangement of time for drawing is very varied, ranging from ten periods a week for a term to a double period a week throughout the four years. In eight cases drawing is continued through two or more years of the course, in six cases it is a part of the first year's work. Freehand, geometrical, mechanical drawing are the common designations. One sequence runs thus, freehand drawing, geometrical drawing, mechanical drawing, building plans.

In only one case does drawing seem to be given for the sake of drawing. Correlation with shop work is more or less close, more markedly so in the agricultural schools than in the high schools. Drawing, where not a special course, is regularly given with the shop work, in some cases the very first drawing being the attempt at working drawings for shop projects. Indeed, in two cases the shop course seemed virtually a course in the construction and reading of working drawings rather than in the use of tools. Said one teacher, "I emphasize the drawing and mathematical side. The boys can't be made workmen here." In poultry, animal husbandry, and farm management, some drawing is often given in connection with building and farm plans.

AGRICULTURAL SUBJECTS

ELEMENTARY OR GENERAL AGRICULTURE. *Table 18*

In all schools where but a single unit or half unit of recognized Agriculture was given, the subject is a text-book survey of the productive fields of agronomy and animal husbandry. Outdoor and laboratory work play but a small part. The aim is 'prevocational' if any. Warren's "Elements of Agriculture" and Mann's "Beginnings in Agriculture," among the books that give at best but a sketchy and inadequate treatment to the productive activity of farm life, are favorite texts. But in four-year courses in which the work becomes more intensive and diversified, the introductory course is still sometimes identical with that above. Even in the agricultural schools in which the particular divisions of the economic subjects prevalent in the agricultural colleges have been handed down from above, the same subject appears four times. The purpose here is the giving of 'back-ground', 'viewpoint', or 'conception of the whole', in order that the subsequent divisions may be seen in their proper relations to one another. Even in project schools, where the elementary survey is not countenanced by central authority, principals report that they have felt it necessary to retain it in order that the specific instances of production may not remain isolated in the minds of pupils, and that the science of production from the soil may have a comprehensive unity. The arguments are plausible, at least, though they seem to presuppose the necessity for deductive treatment. In practice the subject is perhaps the most academic, the experience involved the most uniformly vicarious, to be found in the whole range of the productive topics. If a student is to gain his point of view, his sense of unity, his apperceptive basis for more specialized work, in a series of lessons remote from the realities of participation, then the first-year survey is justified. On the other hand, if, inductively, through the correlation of specific typical experiences, the unity and interrelation of scientific principles is to be attained, then the unifying survey, made meaningful by previous experience, would seem to belong at the end, rather than at the beginning of the course. Such a survey is made frequently in the more or less academic work of Farm Management in the fourth year.

But several principals have advanced the argument for general agriculture in the first year, that it is a provision for the student who does not go on to graduation. If he is to leave school with that most important bequest, the proper viewpoint, then it must be given him at once. The question, then, becomes one as to whether the course given does create the viewpoint. In the mind of the writer it is very doubtful that it does so. The question as to the relative values of the academic survey and of the particular selected experiences of students, even in a narrow field, if only the one or the other can be given, remains open. Only results as determined in the future life of students, if by any means we shall ever succeed in eliminating factors educational outside of the school, can settle it. Such results are nowhere available for study.

The Common Divisions of Agriculture in the Four-Year High School Course

From examination of proposed four-year courses of study put forward by state educational authorities or agricultural college professors, in seventeen states, the figures of Table 19 are put together. The nine divisions of subject matter there listed do not allow of any sure generalizations, but combined with results in the following analysis may be, perhaps, resolved into a sequence like this as a type:

Year I	*Year II*	*Year III*	*Year IV*
Soils Vegetable Gardening	Farm Crops Fruit Growing	Animal Husbandry Dairying Poultry	Farm Management Farm Mechanics

That the suggested outlines dominate the organization actually in the schools is not absolutely proved, though the indications are strong that they do so. Under the control given central authorities by the awarding of state aid such dominance is marked. In the state supported high schools, as in Alabama, it is absolute, and deviation to meet special seasonal, or pedagogical needs is difficult. Elasticity is not a marked feature.

Agronomy. *Table 20*

The relations of soils and crops as a first- or second-year subject in the high schools appears fourteen times, and twice in the agricultural schools. The name is an inclusive one, the material studied being largely that in the courses specified Soils, or Soil Management, and Farm Crops or Field Crops. That the word has not yet acquired a definite meaning seems to be the case. Agronomy and Soils occur in the same course, and Agronomy and Farm Crops in the same course. In the one case Agronomy is the study of crops, in the other the study of soils. The more specific title Soils and Crops is preferable as carrying meaning to ordinary ears, and indicating even to teachers, the nature of the course. The designation of Soils twenty-two times as a separate topic and of Farm Crops twenty-one times, is a move toward classification. As the monstrous Zootechny has been superseded by the plain English Animal Husbandry, so may the Greek give way to the English in this case without loss of dignity.

Soils. *Table 20*

Under the term Soils the matters of origins, physics, tillage, drainage, irrigation, and manuring, are commonly studied. All these topics, save irrigation, appear as special subjects in agricultural schools. Drainage is occasionally taught under agricultural engineering or farm management. As Soils is handed down from the colleges, it is, in many of its phases, one of the most technical and abstract of agricultural subjects. The effect of college presentation is evident in much of the school work, wherein teachers are attempting to give to students void of science training, that which in college is of more than ordinary technological difficulty. In colleges, chemistry is looked upon as a prerequisite to the study of soils, but, as has been noted, such preparation is not provided for in these schools. A few schools, in recognition of the 'needs of the subject' have placed Soils in the last year of the course rather than the first. Two or three have done so primarily for pedagogical reasons that will be noted later on.

It is fair to say that there is in several cases a definite attempt at selection for first-year students. Fortunately, this selection tends to the emphasis of the physical aspects that are, after all, controlling.

Taking all schools together Soils is normally a half-unit course, though in the high schools the full unit is more common. The differentiation of the subject, particularly in the separate treatment of manuring, drainage, etc., probably accounts for the apparently lighter emphasis in the special schools. The shorter course is a factor also. Organization according to season is not marked, or need it be in case of Soils, provided the course carries over a term of open weather in which outdoor work may be done.

Farm Crops. *Table 18*

Season is an important matter in any adequate study of the crops of the farm. To study crops in the field outdoors is far better than to study them only in book and laboratory. In recognition of the value of seasonal adjustment Farm Crops as a half unit is generally given in the fall. As a whole unit it runs through the year or begins in the spring and is taken up again in the fall. Thus a study of the plant at seed-time and at harvest becomes possible.

Farm Crops appears an equal number of times as a first year and as a second year subject. Under the plan of alternation it may share with Soils as the common introductory subject of the agricultural course.

The term Farm Crops is inclusive. It may be made to cover all plant control through agriculture, and it does, in some cases, include vegetable gardening, and even the woodlot. But, commonly, the orchard and small fruit crops and the woodlot are given separate treatment. The usual content is a study of the cereals, grasses, legumes, and root and fiber crops. Selection in view of local conditions is not as usual as it should be, but is fairly common. The Northern school boy usually studies cotton, though he does not give much time to it, the New Englander may study wheat, and the Alabamian, oats at some length. But selection is more or less defined in many cases.

No figures were obtained on the visits to show the amount of time given to various topics in 'agronomy', but inquiry as to the time given to corn, where answers were definite, showed a wide variation. The results of an inquiry by Professor Lusk among schools of the single state of Minnesota show a surprising range of variation. (See Table 21.) To corn was devoted from one to thirty weeks, with a median allotment of eight weeks, to small grains from one to sixteen weeks with a median of eight weeks, to legumes from one to twelve weeks with a median of four weeks, to root and fiber crops from one to seven weeks with a median of two weeks, to weeds from one to sixteen weeks, median three weeks, to grasses one to eighteen weeks, median three weeks, to soils and rotations one to twenty-eight weeks, median four weeks. For corn, the highest figure given in answer to the question by the writer was twenty-two weeks, the lowest four weeks. That standardization of time allotment is desirable is open to question, but that such extremes of variation can be justified by ordinary local specialization, is doubtful.

HORTICULTURE. *See Table 18*

Horticulture is an inclusive term for the study of certain rather arbitrarily selected crops of the farm, usually grown under a rather more intensive culture and on smaller areas than the staple crops. Apples, which are a staple crop, often grown on large areas, and by no means intensive methods, are, by common consent and tradition included among the crops of the garden. Sugar beets and 'mangels' are studied under Farm Crops, table beets under Horticulture, potatoes either under Farm Crops or Horticulture, dent and flint corns under farm crops, sweet corn and beans under either caption or both. Though the degree of intensity of cultivation largely governs selection, the line between Farm Crops and Horticulture is not determinate. In general, two groups of crops are treated under Horticulture, the vegetables, and the fruits. As an undifferentiated subject this general Horticulture is given as a half unit or less in fifteen schools, most commonly in the spring, but in six cases running through the school year.

VEGETABLE GARDENING AND FRUIT CULTURE. *Table 22*

More common, however, than the inclusive caption is the division of the intensive crops into Vegetable Gardening and Fruit Growing, frequently designated Pomology. Under the derived name such small fruits as strawberries, raspberries, blackberries, etc., grouped as small fruits and bush fruits have no place, but they are treated sometimes with Vegetable Gardening, more frequently with Fruit Growing. Vegetable Gardening is commonly a first or second year subject making half a unit of work. Seasonal adaptation, as is fitting, is marked, the spring being the time chosen.

Fruit Growing commonly emphasizes the orchard fruits, notably the apple. The favorite year is the third, the favorite allowance one-half unit, and the favorite season autumn, when varieties can be studied. Occasionally, as with Horticulture and Farm Crops, the cycle of growth is followed by making the beginning of the work in the spring, the end in the fall.

Further splitting up of Horticulture is common in the agricultural schools as in the colleges.

The study of individual crops follows a fairly uniform plan, the treatment of topics running something as follows: Botanical nature of the crop, soil and climatic adaptations, varieties, methods of propagation, culture, enemies, harvesting, storage, and marketing. Emphasis is varied, but in general the control factors dominate. It is, perhaps, fair to say, that with the exception of fruits, the marketing side is slighted.

Animal Husbandry and Dairying. *Table 23*

The two subjects of Animal Husbandry and Dairying commonly go together, Animal Husbandry being the inclusive term. Where it is given alone a variable treatment of dairying is made, and the work includes a brief study of Poultry. When the two are separated the Animal Husbandry usually treats of the types and breeds of farm animals; with the care and feeding of each, and a variable amount, usually small, of stock judging. The treatment of breeding is necessarily brief, if it be heeded at all. When the two are given separately a considerable emphasis is given to the study of milk and its products. For instance, Wing's "Milk and Its Products" is a more frequent guide in dairying than Eckle's "Dairy Cattle" or other books on dairy farming. In such cases the breeds, feeding, and management of dairy cattle are likely to be taken up in the Animal Husbandry course.

Poultry. *Table 23*

The types and breeds of poultry, their adaptations, housing, care and feeding, hatching, rearing, breeding, and marketing, constitute a separate division of Animal Husbandry in fifteen high schools and seven of the agricultural schools. Practical work is more frequent here than in any course, barring shop work, except Vegetable Gardening, with which it shares as the introductory work of the project schools. The frequency of its appearance in the first year is attributable to the New York outline, but by alternation it becomes frequently a second year subject. It is interesting to note that Poultry has been chosen as the introductory topic, in this way, whereas, in general, Animal Husbandry goes into the second or third year, and Dairying into the third year. Discussion will be given to the adaptation further on. See Discussion of General Applications.

Still further division of Animal Husbandry is noted in the agricultural schools.

Reference again to the data of Professor Lusk shows a lack of uniformity in time devoted to the topics of the general Animal Husbandry course. As in the Agronomy it is to be doubted that local specialization in a single state will justify so great a range in intensity of treatment. The distribution is interesting in view of the objection to the Stimson plan of projects as tending to too great specialization. The work in Minnesota is not under the Stimson plan. To that we shall advert in the discussion of the project method.

The time devoted to study of the horse ranges from one to eighteen weeks, with a median of five weeks; of dairy cattle, particularly important in Minnesota, from two to twenty-four weeks, with a median of four weeks; of dairy products and management from one to twenty-four weeks, with a median of four weeks; of beef cattle from one to nine weeks, median four weeks; hogs one to eight weeks, median four weeks; of sheep one to eighteen weeks, median three weeks, of poultry one to sixteen weeks, median three weeks; of feeding one to eighteen weeks, median four weeks. A normal course of thirty-six weeks would thus give to horses five weeks, to dairy farming eight to ten weeks, to beef cattle four weeks, to hogs three to four weeks, to sheep two to four weeks, to poultry two to four weeks, to feeding four to five weeks. Such an arrangement is not far from that called for by the ordinary textbook.

Farm Management and Rural Economics. *Table 24*

Evaluated in terms of thought content, the course in Farm Management ordinarily stands first among the productive agricultural subjects. Its problems, though for the most part vicarious and lacking in immediate motive, most frequently call for the arraying, selection, synthesis and application of facts and principles already more or less familiar to the student. It serves to correlate and organize the various factors in production, and to give that unity, the seeking of which is the excuse for the interpolation of elementary

General Agriculture. Most of all, it serves to the fixation of the economic point of view, whereby science becomes a means rather than an end. The interrelations of land, capital, and labor, not in general, but in more or less concrete instances, form the subject matter. Thus the place of Farm Management is in the last year of the course.

When it is placed earlier, it tends to degenerate into an informational subject on a par with General Agriculture or to become a more or less systematic course in fictitious accounting. This, even in its proper place, it too frequently becomes, particularly in the agricultural schools. For the content of Farm Management is not yet fully determinate, as a survey of publications will reveal. Since the publication of Dr. Warren's book in 1913, the tendency has been to erect the subject into a course, both in schools and colleges, that is really worth while.

Not all the high schools, even those carrying four units or more, teach Farm Management, and two of the special schools omit the division. The common reason assigned for the omission is that: "We do not need it. We teach farm management all the way along in connection with the other subjects." But insisting upon the application of principles to farm practice in all subjects is not teaching Farm Management. Nor did it appear that in schools omitting the subject the concreteness and applicability of the various subjects was one whit more meritorious than in schools where the distinction exists.

Rural or Farm Economics tends to overlap the Farm Management course. It is the more inclusive subject, dealing with economic factors in their wider aspects, and tending to greater remoteness than the Farm Management courses. Land, labor, capital, transportation, marketing, cooperation, accounting, all may enter. But they enter as factors determining the social life and institutions of the country rather than of the individual farm. The larger aspect may not be neglected in the Farm Management course, the smaller may be insisted upon in the Rural Economics. It is not always possible to tell upon entering a school room whether Farm Management or Rural Economics is the general subject. Some Farm Management teachers make use of Carver and Taylor; at least one teacher of Rural Economics makes Warren his guide.

Farm Mechanics and Agricultural Engineering. *Table 24*

In general, divisions of subject matter under the captions: Farm Mechanics, Farm Machinery, and Agricultural Engineering have been grouped together. The caption Farm Mechanics is particularly without definition. The writer has seen fit to list certain of the offerings under that caption with the woodwork and forge work, when the material studied seemed to indicate that the conception of Farm Mechanics was identical with that of the shop courses. The making of farm conveniences such as knots, halters, and ties of rope, fences, gates, bridges, farm and barn plans, may be included in the mechanics course. More often the meat of it is found in the study of the physics and uses of farm machines and powers. One offering might well have been called 'The Gas Engine'. Sanitary piping and sanitary appliances are included, and as Agricultural Engineering, surveying and drainage. Thus the work may have a wide or a narrow scope as is the case with the shop work. Certain high schools, where shop work is offered, consider the special division unnecessary; among schools with farms the work of the farm or the shop is considered adequate to the treatment of the material, and in some cases it doubtless is, just as in some schools where no shop work, so called, appears in the agricultural course, the outside work calls for more of construction and repair than in others where the work is assigned a definite place, so, when the farm work is real, a more adequate knowledge of machinery and its uses may be obtained than in some formal courses. But Farm Mechanics is designated separately in eighteen high schools as a whole or half unit, and in nine of the special schools. It is commonly placed near the end of the course, but may extend through two or three years in the special schools.

Special Topics in Agriculture

The greater specialization and topical division of the agricultural subjects in the agricultural schools than in the high schools is shown in the Tables 25, 26.

Feeds and Feeding as a special subject appears as a half unit in three Congressional district schools, in seven of the ten special schools, and in the preparatory school. In the special schools, it appears in the last year of the course.

Types and Breeds is somewhat less common in the same schools, usually preceding the Feeding. As one-fourth unit or less, these specialized topics may not represent a much more intensive treatment than is found in the high school division in Animal Husbandry.

With the topics of Breeding and Stock Judging, the treatment is undoubtedly more complete than in the high schools. The possession of live stock gives to the Stock Judging in particular, a more satisfactory basis than is likely to be found by even the more ambitious teachers of the high schools who make use of neighboring farms. Criticism is naturally more free, and the correlation of performance with the placings can be more readily followed up.

Some study of barn sanitation, hygiene of animals, and common diseases, may make a part of the Animal Husbandry course. As Veterinary Science a special division is set off in eight agricultural schools, always as one-fourth unit or less.

Swine Husbandry as one-half unit and The Horse as one-fourth unit appear in a state school. Meats in one state and one philanthropic; Live Stock Practice, Judging and Feeding, in one county agricultural and one state school. Animal Management in one state and one philanthropic school, Dairy Manufactures in the same schools, and a study of the broader problems of nurture and heredity in a division called Domestic Plants and Animals in a philanthropic school. Special divisions in Poultry have already been noted in the description of the state school.

Ornamental and Landscape Gardening, ordinarily slighted in the Horticulture course, is definitely treated in seven schools, usually in the spring. In two cases, it is an intensified treatment of a full unit's value. The two high schools in which the division occurs are suburban project schools.

Forestry is a required division in the state-aided schools of Pennsylvania, and appears as a noteworthy adaptation to local needs in two New England high schools. The placing in first and third years is equal, the choice of spring as the season clear, the allowance one-half unit or less, except in the case of the academy previously described.

The study of insects and diseases is in the high schools usually taken up along with the particular crops which they attack, but Insects may become a special topic in the agricultural schools. Plant Pathology is given as two and one-half units and one-fourth unit in the second year by one philanthropic and one state school. Weeds appears as a one-fourth unit in a county agricultural school,

Bacteriology as one-fourth unit in one state and two philanthropic schools, Sprays and Spraying in a state school.

Greenhouse Management is given briefly in one state school and as a half-unit in a philanthropic school. Plant Propagation or Nursery Work appears in one country agricultural school and one philanthropic school, Small Fruits as a spring half-unit in a state school and as a full unit in one country agricultural school, and Floriculture as a whole, a half, and a quarter unit respectively, in one state, county, and philanthropic school. Bee-keeping is given in three agricultural schools, one of each sub-group, and is touched upon in the Animal Husbandry of at least three high schools.

In Farm Crops, Cereals becomes a special division in one philanthropic and one county agricultural school, Forage Crops in one Congressional district school, Root and Fiber Crops in one county agricultural school.

Soil Management is differentiated from the general Soils course in one county agricultural school, Soil Physics in two. But the most frequent differentiation is in the case of Manures and Fertilizers. The division is not here restricted to one group of schools. (See Table 20.) The choice of spring as the proper season is, perhaps, worth noting, though it seems to have little or no bearing on method.

Undifferentiated Political Economy of an elementary sort is given for a term in one of the county agricultural schools. Rural Law as a half-course appears in the last year of two town high schools. The study of rural institutions is made a half unit in the last year of the course in one academy and one philanthropic agricultural school.

On the whole, the study of social science, particularly in its country aspects, is very much neglected. Such work as is included in Farm Management in some schools is negligible. It is ordinarily based on the brief chapters in Warren's text.

The omission of this study merely serves to reflect the emphasis upon production, which is a characteristic of the agricultural course. Though economic independence is largely basic to the betterment of country living, success in production alone will not solve the rural life problem. No need of the country is more marked than that for constructive leadership and an attitude of sympathetic cooperation with its undertakings in the social field. An increasing definition and coordination of institutional activities and collective effort constitutes social progress. That such progress in church, school, grange, club, cooperative association is to come in the absence of an intelligent appreciation of the problem involved, is hardly thinkable. Some little may be done by good teaching in respect to these matters to affect attitudes and sympathies among those who are preparing to be successful producers in the coming generation. Already, Rural Sociology has established itself as a division of social science sufficient in thought content and usable values to justify the incorporation of its fundamentals in the course of study of any school that pretends to prepare for country living. No part of the course offers better opportunities for the development of the 'sense of kind'. Indeed, so commonly is social contact a felt need of country life, and its satisfaction a first motive to effort, that it is remarkable that even so isolated an institution as the school can have failed to seize upon and use it. When the nature of the child and his home rather than 'the nature of the subject' shall control, such omissions will be more rare.

SHOP WORK. *See Table 27*

Perhaps the most difficult of all subjects to classify are the divisions in Carpentry and Forging. In some schools, where no regular course is given, the practical value of work given in repairs and construction on the farm and about the buildings is greater than in most of the formal courses. This is notably true in the case of the philanthropic schools and of the outside work of the Congressional district schools, in only one of which is the formal shop work in any wise adequate. Woodwork is much more common than Forge Work in the high schools, being definitely set off in twenty-eight of them, as against Forge Work in nine. Five times the work appears as undifferentiated formal process Manual Training for agricultural and other students alike. In the other cases work is very variable, as will be noted under Method. It is usually first-year work, but may be extended through the full course. In general, the agricultural schools with their more complete equipment have developed more elaborate courses in which the outcome is a considerable attainment of skill. That their work is more closely adapted to the farmer's needs, it is, perhaps, unsafe to say. Schools under the project plan usually teach some woodworking, whether it be a recognized course or not. Seven schools of Group B carry Woodwork and six Forge Work, the content of which is by no means uniform or determinate in type. A sample of organization of the more complete sort is here given. Further discussion is reserved.

COURSES IN SHOP WORK IN A PHILANTHROPIC SCHOOL

Year I		*Year II*		*Year III*	
Fall		Farm Repair Shop	6	Farm Construction	4
Winter Farm Shop	6	Farm Repair Shop	6	Farm Blacksmithing	5
Spring Farm Shop	3	Farm Repair Shop	6	Farm Repairs	3
				Woodwork and Blacksmithing	3

That woodwork commonly precedes forge work, and that the type unit is one-half appears in the table.

Concrete Construction appears as a definite subject in three county agricultural schools, in one case involving a full half unit. The inclusion of the work in concrete under Farm Mechanics, Engineering and Shop in the project schools is quite usual. As a selection in view of the farmer's needs it should stand not far behind carpentry.

ARRANGEMENT OF TIME

In so far as agriculture is to be taught through active participation of the pupil in real experiences, it makes demands for a larger allowance of time in the daily program of the school than other divisions of the curriculum. In that it involves undertakings that cannot be interrupted at the stroke of the bell, it is like cooking or forge work in requiring an extended period. In that many of the most useful activities, particularly in the high school, must be carried on, if carried on at all, at some distance from the schoolroom, the demand for freedom in time may well be even greater than in those studies. Recognition of the necessity for special provision in the matter of time seems to be growing, particularly in the state-aided project high schools. For instance, both the New York and the Massachusetts plans call for longer periods than are given the academic subjects or even the sciences. In the one case a double school period every day, in the other a triple period every day is set apart for each class in productive agriculture. Several such schools have set apart the half day, free from the demands of other teachers upon pupils, to be used by the teacher of agriculture unhindered. In some schools where the single period still prevails, the agricultural classes are scheduled for the end of the school day, so that the work may not be restricted by the necessity of getting back to school in time for the German lesson.

Arrangements of time for agriculture may be noted in Table 28. The common length of the school period is forty minutes. Eight high schools, one preparatory, and two philanthropic schools, hold to the single period arrangement, that is probably the norm, if all schools of the country be considered; five high schools provide four single periods a week, with one double for laboratory; seven, three single and two double laboratory periods, as with other science subjects. This last arrangement is near to those of three state and three county agricultural schools, which provide single periods for lecture or recitation, double for laboratory or outdoor work. All agriculture periods are double in sixteen high schools, all triple in three high schools and two special schools.

Placing of all agricultural work in the forenoon is made in five high schools, better still, in the afternoon in ten. Such adjustment is not so necessary in the special schools located on farms, and may often be impossible because of the greater range of divisions in agriculture and the generally greater proportion of time devoted to it.

Alternation of years, so that two class-groups study together, as has been noted in the description of schools, is the practice of seventeen high schools. In a course with one teacher, involving four or more units of agriculture, or requiring part time of the teacher for other duties, the arrangement becomes a *sine qua non* of success. Schools having two or more teachers of agriculture do not require it. One high school where three units only have ever been elected hires an assistant for the teacher of agriculture every other year, and maintains the printed sequence.

Teaching six days in the week is not common even in the special schools with their boarding pupils. It is reported in three cases only.

METHOD. *See Tables 29–30*

General returns under Method are given for the classroom, the laboratory, the shop, and outdoors, and in the adoption of home and school projects. Such returns are based upon information from principals or teachers, or both, in all schools, supplemented by some reports from students and by observation at the time of visit. The particular reports of illustrative lessons were taken in long-hand by the visitor.

Classroom Method

The use of text-books was regular in some, usually in all of the agricultural subjects, but shop, in forty-five of the fifty schools. Five high schools did not use regular texts, but made the basis of recitation assignment of topic references in books or bulletins. Assignments were made to the class as a whole for recitation, and individually to members of the class for special report. The class recitation as heard under the assignment method did not seem to differ in procedure from that in the regular text classes, but the individual reports appeared to call forth a rather more full and critical discussion by students than in the usual case. Pupils who would not think of disputing the findings of a book did not hesitate to differ with conclusions stated by a fellow. For reports by individuals in the classroom savor of the personality of the reporter, whether original or no, and, because of that, lack the sacred odor of infallibility of the common text. The high school boy who will question a text is rare; the boy who will not question a fellow student is, perhaps, nearly as rare.

Even when the regular text was the basis of classroom work it was supplemented in seventeen of the high schools by topic assignments, commonly in connection with projects. Under the influence of the project, the tendency to break away from the text-book order of topics was common. In that way the text might be made useful in preparation for practical application, or, more rarely, to answer questions already raised by participation of the pupil in some project activity.

The greater number of teachers reported some minor departure from the order of the text; a few made the text subserve their own notions of sequence in topics. It is not unfair to say that the great majority served as media for the interpretation of the text, rather than as teachers who make the function of the text interpretation and supplementation of their own teachings. All too commonly the business of the class was to answer questions suggested to the teacher by the text. Only rarely were text and teacher called upon to answer questions arising in the experience of the pupil. In this the teaching of agriculture does not differ from other high-school teaching.

Recitation

As may be inferred from the foregoing, the question and answer method, by which the teacher attempts to learn whether or no the class has read the assignment or listened to the lecture, prevails in the classroom. Every school reported the use of this method. In forty-two schools it was the major method of procedure in the classroom, if not the sole method. Twenty-five schools used lectures in addition, only two high schools, however, making the lecture dominant. In conformity with their approximation to the college organization, the use of the lecture was common to all special schools, six of the ten making it the usual, and in the view of directors and teachers, the more important method of classroom procedure.

Exhibits of Question and Answer Method

The following exhibits may serve to make more concrete the discussion of classroom method. They do not represent extremes, but a gradation from rather poor teaching to fairly good teaching. As extremes, perhaps, two were outstandingly poor. In one, a recitation in animal husbandry, the teacher took no part, except to direct the method. Every boy opened his book to the assigned page. Then, beginning at the right, a boy stood and read aloud the first paragraph. When he had finished, his next neighbor rose and repeated, as nearly as he could remember, what had been read. And so on through the period. Not a question was raised, except as to pronunciation of words, and the class was too drowsy to give much heed to slips.

In another case the activity was all on the other side. Noticing an extraordinary rapidity of fire on the part of the teacher, the visitor laid out his watch to check off the number of questions. At point blank they numbered 130 in twenty-five minutes. The speed was maintained successfully by the teacher through giving the answers himself. In only three cases did a pupil interject the necessary word before the teacher could utter it.

At the other end stands a discussion of projects by a class, each of whom had chosen and made preparation for an undertaking at home. He had noted difficulties and brought them to the class for help. It was a round table discussion conducted wholly by the boys, at which the teacher sat merely as a companion. Threshing out led to appeal to him for confirmation or rejection of suggestions in two or three cases, but his words were brief.

LESSON IN FARM CROPS

Topic. Insects Injurious to Corn

T. How could you tell a cut worm from a wire worm?
P. The wire worm is more like a wire.
T. Is the body wall of the wire worm, as compared with the cut worm, soft or hard?
P. Hard.
T. How does the body look?
P. Shiny.
T. How about the color of the cut worm?
P. It is kind of a dirty color.
T. Did you find out what they are larva [*sic*] of?
No reply.
T. Well, the wire worm is the larva of the *May beetle.*
T. What is the difference between a butterfly and a moth?
P. I think butterflies are generally more brittle than moths.
T. No, I don't hardly think there is much difference in butterflies. Let's have a look.

He goes out and brings in a case of mounted Lepidoptera, which he holds before the class while he describes the antennæ of moths and butterflies. Then he lays the case on the desk, and draws on the board the clubbed antenna of the moth. The case is thenceforth ignored, no pupil having looked at the insects from a closer range than six feet.

T. What senses are in the antennæ?
P. Touch, hearing, smell, taste.
P. Where are the nervous centers located in moths?
T. In the back. But that is getting too far away from the subject.
T. Where are the eggs of the cut worm and the wire worm laid?
P. In the ground.
T. How long will the larva of the wire worm live in a sod?
P. Two or three years.
T. Why isn't injury noted in grasses and small grains as it is in corn?
P. Because there are more plants and more roots in the soil.

T. How about prevention? For cut worms?
P. You can make a dough of meal and bran and Paris Green, and poison them.
T. Any other way?
P. Yes. You can plow them in, in the fall.
T. You mean you plow them in so deep they never come up, don't you?
P. Yes.
T. You haven't that very good, have you?
T. How about the white grub?
P. It injures corn.
T. What is it the larva of?
P. Of the May Beetle.
T. That's right. It isn't the wire worm. I knew that. But I said it to see if you would notice.
T. What about the life cycle gives us a chance to attack it?
P. You can kill the grubs or you can spray trees and kill the beetles.
T. What effect will what we do for wire worms have on the white grub?
P. Plowing in the fall helps some.
T. Describe the white grub for us.

After several attempts on part of pupils he describes it himself.

T. (*facetiously*) What use is the white grub?
P. It makes good bait.
P. Why isn't it safe to put Paris Green on cabbages?
T. Because the Paris Green *works into* the cabbage leaves and *makes them poisonous.*
T. (*turning page of the text*) How about the root louse? How does the root louse produce injury?

Pupils make numerous absurd guesses, among them, these: They drink up all the water before the corn can get it. They dig the soil away from the roots so the corn dries up. They call ants, and the ants sting the corn to death.

The teacher closed the book, slammed it on the desk, and proceeded to berate the class. Then he gave up the remainder of the period to a talk on the relation of ants and aphids, in which he displayed a knowledge of the subject much more creditable than is revealed in the foregoing report.

This teacher was employed in a state-aided town high school at a salary of $1,400. He was a graduate of an agricultural college, born and raised on a farm.

LESSON IN SOILS

Topic. Nitrogen

T. What are the ways in which Nitrogen is lost from the soil?
P. By leaching, cropping, erosion, and denitrification.
T. How is Nitrogen lost by leaching?
T. What do you mean by leaching?
P. Rain water works into the ground and washes out the Nitrogen.

T. It is lost through drainage, isn't it?
P. Yes.
T. What is the difference between leaching and erosion?
P. Leaching washes it through the soil; erosion carries it off over the surface.
T. How is Nitrogen lost by cropping?
P. When you take off a crop of tobacco and don't put anything back, you lose Nitrogen from the soil.
T. Do weeds take off Nitrogen?
P. Yes, unless you plow them in.
T. In what soils does denitrification usually take place?
P. In poorly drained or poorly tilled soils, where the air can't get in.
T. What are the different sources of soil Nitrogen?
P. Lime.
T. No.
P. Kainit.
T. No.
P. Nitrate of soda, barn-yard manure, sod and stubble.
T. In case you plant a legume, is that a source of Nitrogen?
P. Yes.
T. Where does the Nitrogen come from?
P. The air.
T. Then the sources of Nitrogen are manure and air, aren't they?
P. Yes.
T. Those Soy beans we saw last week have been plowed under. What did they bring to the soil?
P. Nitrogen.
T. What other crops could you plow under to get Nitrogen?
P. Clover, alfalfa, cow-peas.
T. What are some of the commercial sources of Nitrogen?
P. Ammonium sulphate and sodium nitrate.
T. What is another name for sodium nitrate?
P. Nitrate of soda.
P. Chile salt-petre.
T. Where does it come from?
P. Chile.
T. Any other sources?
P. Dried blood.
T. How much nitrate of soda should a farmer apply?
P. Forty to sixty pounds to the acre.
T. What happens if he applies too much?
P. It kills the crop.
P. It makes the soil hard.
P. It makes a kind of plaster.
T. Is nitrate of soda very soluble?
P. No.
P. Yes.
T. What happens to the soil if you apply too much ammonium sulphate?
P. You spoil the soil and kill the crop.

P. It is likely to make the soil acid.
T. What should you apply with the sulphate?
P. Lime.
T. Where do we get dried blood?
T. At the slaughter houses, don't they?
P. Yes.
T. About how much should be applied to the acre?
P. About 200 pounds.
T. Isn't that too much?
P. I don't know.
T. Well, how much mixed fertilizer is usually applied?
P. From 300 to 500 pounds.
T. Are there any bad effects from applying too much?
P. It will spoil the crop.
T. Does it have any bad effects on the soil?
P. No.
T. Well, it *is very slowly available. It may take two years to do any good.*
T. About how many pounds of ammonium sulphate shall we put on an acre?
P. Twenty-five.
P. Forty.
T. (*Advises frequent small applications of nitrate of soda, emphasizing danger of using too much.*)
T. The answers came too slowly today. We shall spend the next two days studying Nitrogen, and then have a test. I want you to pay attention to the five effects of areation, so that if I ask you, you can tell why.

LESSON IN POULTRY. DOUBLE PERIOD

Topic. Appliances

The following appeared on the board:

"Always in this course before recitation outline the assignment before you come to class. Do this in Poultry, Soils and Forestry."

CONTINUATION OF OUTLINE

Poultry Appliances

6. The Broody Coop.
   A. Defined: is a coop designed to hold broody hens.
   B. Essentials:
      a. It must be cool.
      b. It must have plenty of light.
      c. It must be large enough to provide one square foot of floor space for every twenty hens in the laying pens.
         (a) This varies with the breeds.
         (b) Should never be less than three square feet.

7. Dust Wallow.
    A. Use:
        a. Keeps down lice.
        b. Gives some exercise.
        c. Hens like it; it adds comfort.
    B. Location:
        a. Outside of house, so dust will not be in the house.
        b. May be in one corner of house.
    C. Size:
        a. At least four square feet of dust-wallow for every 100 hens.
    D. Materials:
        a. Sifted *coal* ashes.
        b. Ordinary road dust.
8. Shipping crates.
    A. Essential that a crate be:
        a. Light.
        b. Strong.
        c. Airy.
    B. Materials:
        a. Light.
        b. Strong.
        c. Cheap.
9. Chick Shipping Box.
    A. Should be comparatively tight—only air enough to keep chicks going.
    B. Box no higher than the chicks.
    C. Not more than fifty in a box.
10. Dropping Board Platform.
    A. Advantages:
        a. Keeps house much cleaner.
        b. Gets droppings in pure form.
        c. Decreases air space around birds—keeps them warmer in winter.
        d. Decreases drafts on birds.
    B. Disadvantages:
        a. Costs more money.
        b. Requires more labor.
    C. Arrangement: run the boards from front to back.
    D. Cleaning hoe: ordinary broad hoe with board attached.
11. Fattening Coop: used especially for fattening cockerels.
    A. *a*. It should be airy.
        b. Room along front for all birds to eat. Prevents fighting.
        c. Put only ten birds in a coop.
    B. Construction:
        a. Wire netting stretched over board floor.
12. Trolleys.
    A. Make it of barn door track and rollers.
    B. Be sure you want it before you install it.
13. Catching hook.

RECITATION

T. What were we talking about last time?
P. Inside devices.
T. Well, we are going to start some chicks after Christmas. What are you going to have for a chick feeding device?

Every boy had a suggestion or a reason for or against one device or the other. Boys stood and addressed the class, not the teacher.

T. What did you decide was the best watering device?

Discussion indicated agreement on fountains.

T. Why would you rather have an earthen than a galvanized fountain?

Argument again, with most boys favoring the earthenware as more durable and more easily cleaned, and less easily upset. Objections were that the earthenware is more expensive, heavy to handle, and not easily mended when broken.

T. Where did you see a running water system?

One boy describes the system on a plant visited by the class and several describe systems that they have seen.

T. Now, remembering you have to keep under a dollar a hen, what do you think of the barrel method?

Here followed a prolonged and animated argument on the barrel method versus a piping scheme.

T. What feeding devices have we considered?

Boys name several types of hoppers, fetching them out for inspection.

T. What do you think of the dry mash hopper?

Most regarded it as a great convenience, but one criticized it as unsanitary.

T. What are some of the essentials of an egg crate?

Boys list desirable properties.

T. (*Producing patent pasteboard egg crate*) Well, what do you think of this?

Criticism interested and intelligent.

Here the teacher felt it necessary to complete the assignment promptly and turned to the board. The boys sagged into their seats to answer with mechanical promptness, often by literal reading of the blackboard outline, a long series of questions like the following:

T. What do you mean by a broody hen?
P. A hen that wants to hatch chicks.
T. Is it natural for all hens to want to hatch chicks?
P. Yes, but some breeds don't. White Leghorns, for instance.
T. What is a broody coop?
P. A coop designed to hold broody hens.
T. What are its essential features? Etc.

LESSON IN FARM ARITHMETIC

Topic. Commercial Fertilizers

Lesson began with the assignment of eight problems in the text for the following day. Then followed rapid fire questions, such as are reported, with indicated replies, given for the most part promptly and in interested fashion.

T. What fertilizer materials did you see yesterday?
P. Tankage, dried blood, sulphate of ammonia, nitrate of soda, acid phosphate, ground rock, bone, dried fish, muriate of potash, sulphate of potash, kainit, wood ashes, cotton seed meal.
T. What would you be feeding your plants if you put on nitrate of soda?
P. Nitrogen.
T. Why is this a good form of nitrogen to use?
P. Because the plant can use it right away. It is soluble.
T. Do you remember how Dried Blood appeared?
P. It was a fine blackish powder.
T. What does it come from?
P. The slaughter houses in Chicago.
T. Where else?
P. Buffalo and Omaha.
T. Where would you go to get it?
P. To the men who have it for sale.
T. Well, who are the men who have it for sale?
T. Why, where would you get it in your home town?
T. Who has it for sale here?
T. (*Stops to explain the distribution through wholesalers, jobbers, grocers and grain dealers.*)
T. What would you be putting on the land when you add blood?
P. Blood.
P. Nitrogen.
T. What per cent. of blood is nitrogen?
P. About ten per cent.
T. Does it vary considerably?
P. Yes.
T. What else do you remember from the lot yesterday?
P. (*Gives list again.*)
T. Which ones furnish phosphoric acid?
P. Acid phosphate, rock phosphate, bone, fish, tankage.
T. Which ones furnish potash?
P. Muriate of potash, sulphate of potash, wood ashes, kainit, cottonseed meal.
T. Now, what do we call these things?
P. Fertilizers.
T. Every one of them contains quantities of what kinds of materials?
P. Dry materials.
P. Plant foods.
T. What do we call them, then?
P. Plant food carriers.

T. Now, go to the board, David, and give us the shorthand for nitrogen, phosphoric acid, and potash.

P. (*Writes*) N., K_2O, P_2O.

T. Can you give shorthand for nitrate of soda and for ammonia?

P. (*Writes*) Na NO_3, N H_3.

T. What does that shorthand tell you?

P. Nitrate of soda has nitrogen in it, ammonia has nitrogen in it.

T. Anything else?

P. Yes, nitrate of soda has sodium in it and oxygen; ammonia has hydrogen.

T. Now, has the farmer got to think about chemistry?

P. Yes.

T. (*Hangs up a bag of Essex Complete Manure, empty, so that the printed guarantee is plain.*) Now, tell me what that means. (*By means of rather skilful questioning he elicits a very creditable interpretation.*)

T. Now, find out what you buy when you buy a bag of that. (*Boys spend about ten minutes in figuring the pounds of nitrogen, phosphoric acid, and potash in the bag.*)

T. Now, John, you can tell me what you know about that bag. Harry, David, *et al.*

Boys make very creditable summaries.

T. Isn't it fine to be able to read bags, and to know that the fertilizer man can't take your money away from you?

LESSON IN FARM MANAGEMENT

T. What would you do in case of failure of a seeding in a rotation?

P. Plow and seed again.

T. What effect would that have?

P. It would halt the rotation and upset plans.

T. Which would be more difficult, a rotation on equal fields or one on unequal fields?

P. A rotation on unequal fields.

T. Would there be any advantage in rotation under these conditions?

P. Yes. You would stand a better chance of hitting the market somewhere every year.

T. Go to the board and work out a rotation for a 160-acre farm, calling for eighty acres of wheat, forty acres of hay, twenty acres of corn, and twenty acres of potatoes.

The results varied somewhat. Plans were criticized in order by the class in rather shrewd fashion; then a summary criticism in which practical difficulties were emphasized was made by the teacher.

T. Can you have a five-year crop rotation on less than five fields?

P. No.

T. Why not?

P. It is possible to grow more than one crop on a field in a season, but at least one crop must be grown.

T. Work this out: On a seventy-five acre farm, plan to take off twenty-five acres in buckwheat, twenty-five acres potatoes, twenty-five acres of oats, and twenty-five acres of clover every year.

A pupil's plan:

1915	*1916*	*1917*
Clover Buckwheat	Potatoes	Oats
Oats	Clover Buckwheat	Potatoes
Potatoes	Oats	Clover Buckwheat

Part of Criticism:

T. How do you provide for organic matter?
P. By plowing under clover.
T. Would you plow under a good crop of clover?
P. No. I would cut the first crop and plow under the second.
T. Where would you get in your buckwheat under that scheme?
P. I couldn't get it unless I followed oats.
T. When would you seed the clover?
P. I might seed in the spring but that would not be so good.
T. How often do you remember cutting a second crop of clover round here?
P. It is not very often that you can.
T. Can you carry two rotations on the same farm?
P. Yes.
T. Give an illustration.
P. Corn, oats, wheat, hay and beans, wheat, clover, potatoes.
T. What advantages in wheat after beans?
P. The land is clean, you have time to get in the wheat after harvest, and the beans bring some nitrogen to the soil.
T. Work this out:

A New York farmer has 100 acres. Figure a rotation growing five acres of potatoes, five acres of beets, five acres of corn, twenty acres of oats, twenty acres of wheat, twenty acres of clover, twenty acres of timothy.

T. How many fields do you need for that rotation?
P. Five.

Plans again subject to class criticism.

T. What objection is there to putting wheat on grass land?
P. The grass feeds in the same way as the wheat. Then sod is a little loose for wheat.
T. What are some of the benefits of a rotation?

Pupil recites text list in order.

T. What relation does a cropping system have to a feeding system?
P. Don't know what you mean.
T. Think it over a minute.
P. You have got to provide for forage in planning your rotation, and what you feed will make a difference in the value of the manure.
T. Why is it we have so many one-crop systems in this country?
P. Because the country is new, and one crop usually pays better than any other in the different states.
T. What is the principal crop of the South?
P. Cotton.
T. Why?
P. It pays best.
T. What is the principal crop of New York?
P. Hay.
T. Why?
P. It pays most for labor.
T. Of Minnesota?
P. Wheat.
T. Why?
P. Same reason.

Examinations

No data were gathered with respect to the frequency of examinations, 'quizzes', or 'tests'. Term or quarterly examinations seemed to be the rule. In addition, a number of teachers stated that it was their custom to have a 'quiz' at the 'completion' of every topic. Several lecturers said 'every two weeks', and at least one 'every week'. That is, regular examinations were held according to the rules of the school, additional examinations in accord with the notions of the individual teacher.

Whether the examination be looked upon as a means whereby the teacher may check his own work or the work of his pupils, or as a method of teaching; whether it be a test of memory or a means to organic thought, it may reasonably be expected to show forth that content which, in the mind of the teacher, is of most importance. That is, it reveals a selection of material more discriminating than that of the ordinary recitation. Further, if it be a means to teaching, it should reveal the ability of the teacher to stimulate thinking, to question.

Several specimens of examinations are submitted for study. The form of the question is relatively easy to judge, though in the absence of knowledge of teaching developments in the respective classes, it is not possible to give an absolute value to any question, with regard to its thought-producing properties. On the other hand, it is difficult to precipitate the criteria of selection, if definite standards do exist. In topics dealing with productive agriculture it seems reasonable to judge that one such criterion should be *control.* Knowledge on the productive side may be evaluated in terms of its usability in control of racial and individual growth in plants and animals. Conventional, cultural, and preparatory evaluation may also be made. That definite standards for the selection of material, other than that of reproducing a skeletal replica of the text, do prevail, appears doubtful. In so far as text selection is just, and no such selection can always be just in varying localities and groups, the class is fortunate.

EXAMINATION IN FARM CROPS

1. Give the Latin name for oats.

   Give the names and pedigrees of Wisconsin oats.
   Name as many other varieties as possible.

2. Name the five leading oat-producing states.

   What place does Wisconsin occupy?
   What part of the world's supply does the United States produce?

3. Give the time, rate, and best method of seeding oats.
4. Should we manure oat ground? If so, when, and how much?
5. Give all the uses of oats.

BOYS' PAPERS IN EXAMINATION ON FARM CROPS

A

1. The Latin name for oats is *Venus Setivia.*

   The pedigree names for oats are Wisconsin wonder or pedigree no. 1. Swedish select, or pedigree no. 5; and the sixty-day oats. Other varieties are: rust-proof.

2. The five leading oat-producing states are Iowa, Illinois, Wisconsin, Nebraska and Minnesota. Wisconsin ranks third in amount of oats raised.

   The United States produces about one-half of the total amount produced in the world.

3. The time of seeding oats around here is the latter part of April.
   The rate is two to three bushels to the acre.
   The best method of seeding oats is with the drill which gets the oats in all about the same depth and more evenly sown. Sometimes it is sown broadcast, which is a poor method because the oats are sown too unevenly.
4. Oat ground should not be manured the same year that it is planted except with the crop residues because the ground is too rich and the oat will lodge.
   The crop planted previous to oats should be manured but not the oat ground.
5. The uses of oats are mostly for horses or working animals because it is a muscle-producing food; it is also used for human food and for chickens and mixed with swill for hogs. It is also good for all kinds of growing animals.
   The straw is used for bedding although it is a very good roughage.

B

1. Swedish select or pedigree No. 5.

   Pedigree No. 1 or Wisconsin Wonder.
   Sixty-day oat not pedigree. Other varieties are Big Four.

2. Wisconsin, Minnesota, Illinois, Iowa and Missouri.

   Wisconsin has first place.
   United States produced about all the oats, or two-thirds.

3. The time to plant is the last half of April or early, if good year; later, if poor year; two to three bushels should be planted the acre. The best method of seeds is driller and checker.
4. We should not manure oat ground unless it is necessary, because it makes the soil too rich. It should be manured a year before, or if the manure should be well rotted when put on the oat land, very little should be put on an oat field.
5. Oats are used as food for working horses, growing horses, dairy cows and pigs. When ground, for food for people, as oat meal.

EXAMINATION IN SOILS

1. Name in order of importance the chemical elements necessary for plant growth. Which are apt to become exhausted?
2. What is meant by osmosis and what is its relation to plant growth?
3. Name the conditions for germination and for plant growth.
4. Define the term soil. What is meant by the physical make-up of the soil? Describe a method of soil analysis.
5. Name three classes of rocks and tell what is meant by the names.
   What is meant by a rock-forming mineral?
   Name four different classes of soils with respect to origin.
6. How are soil and subsoil distinguished?
7. What is nitrification? What conditions are necessary to the process?
8. What is meant by nitrogen fixation?
9. What is meant by soil inoculation?

EXAMINATION IN AGRONOMY

1. Does each element which we eat have a duty to perform in the body?
2. Does each element taken up by the plant have a duty to perform for the plant?
3. Can one element be substituted for another?
4. Which contribute most towards the material of the plant, air- or soil-derived elements?
5. Name four things which make nitrogen of so much interest agriculturally.
6. What is the effect of nitrogen upon foliage?
7. What crops benefit by the use of large amounts of nitrogen?
8. What is the effect of nitrogen upon the flowering process?
9. What is the effect of nitrogen upon the maturing of plants?
10. What is the effect of nitrogen upon the color of plants?
11. What is the effect of nitrogen upon the resisting power of plants?
12. What is the effect of nitrogen upon general all-round growth?
13. What is the effect of phosphorus upon the germination of the seed?
14. What is the effect of phosphorus upon the maturing of a crop?
15. What is the effect of phosphorus upon the relative amounts of straw and grain?
16. What is the effect of potassium upon the formation of carbohydrates?
17. What is the effect of potassium upon stems and leaves?
18. What is the effect of potassium upon resistance to disease?
19. What is the effect of potassium upon leguminous crops?
20. What is the effect of calcium upon the root-hairs?
21. What is the relation of sulphur to plant proteins?
22. What is the action of carbon in plants?
23. What is the action of oxygen in plants?

EXAMINATION IN POULTRY

1. 25 wts. Define breed, variety, strain, single comb, rose comb, cock, cockerel, hen, pullet, barring, penciling, squirrel-tail.
2. 50 wts. *a.* Starting with head, name the parts of the hen's body.
   b. Describe the characteristics of the parts you associate with vigor and egg type.
3. 25 wts. *a.* From what country and how were the Leghorns brought to America?
   b. They originally came from what type of native fowl now found in India?
   c. Name five economic qualities of the Leghorn fowl that *has* made it popular in America.

EXAMINATION IN POULTRY

1. Draw a plan for a poultry house for 100 hens.
2. Make a bill of lumber for the house.
3. Draw a sketch of a hen and indicate the parts of the body.
4. What points would guide you in selecting fowl for egg production?

EXAMINATION IN ELEMENTARY AGRICULTURE

1. (10) Tell the uses the plant makes of:
   a. Roots. *b.* Leaves. *c.* Flowers.
2. (10) What are the three parts to a seed, and tell the use of each.
3. (10) Where should a farmer get most of his seed for planting?
4. (10) In sowing wheat, why is drilling preferred to broadcasting?
5. (10) What are four conditions necessary to secure a good crop of corn?
6. (10) Why discard the tips and butts in preparing seed corn?
7. (10) Explain fully how a white kernel may appear on a yellow ear.
8. (10) Why is a cylindrical ear of corn better than a tapering one?
9. (10) How should corn be selected and stored?
10. (10) How should seed corn be tested?

Lectures

The lecture method of presentation of subject matter is common to thirty-five of the fifty schools. It seems to be a direct consequence of the agricultural college training of teachers. Two directors in special schools reported that thus far all efforts on their part to eliminate lecturing had been unsuccessful. The teachers felt that the one way in which they could unify and make clear the various topics of the agricultural course was by lecture. Consequently a repetition, in more or less perfect form, of the instruction of their college days to their adolescent pupils, was frequent. The purpose seemed to be, commonly, not the imparting of information that is unavailable in text or reference, but a formal assembling of facts.

However, the lectures heard were very variable in form and in interest. There was the slow dictation from the typewritten page, the formal elaboration, head by head, of carefully systematic notes, the inspirational sermon, the off-hand and often unorganized discourse. So far as the judgment of the visitor may be trusted, these last seemed of most interest and value. Perhaps the reason lay in the type of men who undertook such a proceeding. These were the men who knew the subject first-hand, who constantly cited illustrations of a concrete nature, who, if not always providing new information, yet reviewed from a new angle the material to be found in books. The less experienced men were, on the whole, apparently more careful and less interesting, for the reason that they recited rather what they had heard or read, than what they had seen or done. In their formal presentation, too, they were less tolerant of and less subject to questions and interruptions by the class. The greater concreteness and the personal touch of the less formal 'talk' led to more frequent questioning and even argument.

The knowledge of the subject matter displayed in the lectures was, on the whole, very creditable. Only one man among those listened to, ventured to expatiate at length upon what he did not know. He preached a sermon upon the evolution of plants and animals that was amazing in its vigor and inaccuracy. In the special schools several of the lectures were of a first-class order. Yet for the most part, the activity of pupils in these classes consisted in scribbling notes at top speed.

Supervised Study

Provision for supervised study in the class period was made definite in seven schools. Just what attempt is ordinarily made to help students to study in this period, it is hard to say. One teacher spent the time set apart for study in completing his lecture, another in a continuation of questions upon the text. Two sat down at their desks and left the class to its own devices. One assigned the writing of a brief composition upon 'Farm Yard Manure', and gave his time to aiding in organization. He moved from boy to boy, asking "Now, what are the heads you are going to make? Which do you think are most important?" etc. Then, as sheets were completed, he read them critically, saying, "Now, why do you say that? Couldn't you state that a little more clearly?" and the like. Two others distributed question outlines for projects, and stood by to give aid in explanation. Both were kept busy by the boys.

The Round-up

A practice, reminiscent of the agricultural college, and suggestive of the farmers' institute, was reported in some half dozen schools. This is the setting apart of a period weekly, fortnightly, or monthly for report and discussion of topics treated in farm papers or other publications. The possibilities for constructive organization and real teaching are considerable under the plan. But the work seen in the two cases of such meetings attended by the visitor was distinctly disappointing. In neither case was there any central topic for discussion, any problem to be solved. The boys had chosen haphazard from publications in the library, and reported verbally, or read, the contents of the article of their choice. In neither case did any serious discussion arise, nor was any original contribution put forward. But a more sustained effort at self-expression than is usual in the classroom, and an actual contact with new sources of information, did result.

The haphazard nature of the program is suggested by the following:

TOPICS REPORTED ON AT A 'ROUND-UP'

1. Rainfall.
2. Silos.
3. Poem, 'The Lazy Farmer'.
4. Rabbits as an Orchard Pest.
5. The Oregon Agricultural College.
6. The Production of 'Hot House Lambs'.

LABORATORY METHODS. *See Table 29*

In the laboratory in Soils, Agronomy, or Farm Crops, the following of a series of set experiments by direction of a printed manual or typewritten outline was the usual procedure. In most cases, students worked as individuals, each one responsible for the doing and recording of the whole number of experiments. That is, the plan of procedure is that usual in the physics and chemistry of the high school. In a few cases, poverty of equipment was such that no ingenuity in the rotation of apparatus could give individuals a chance, and the class worked in groups. Demonstration by the teacher, and in one case by selected students, before the class, is the only method used in the laboratory of six schools. Six high schools did no laboratory work.

Perhaps the most common piece of apparatus in the schools is a hand-power Babcock tester. The presence of that useful little machine is indicative of the dairy work of the high schools. The restriction of dairy laboratory work to milk testing in such schools is almost inevitable. Even the possession of live stock may not justify the use of milk for practice in the making of butter, or more rarely of cheese, though pasteurization might well be more frequently taught, even in high schools owning no dairy cattle.

The amount of time given to testing milk is very variable. Sometimes merely a test by each boy, or even by the teacher, makes up all the work. More frequently, however, the work is motivated by testing of the pupils' home herd or individual cows. Sometimes the high school becomes the medium for testing of herds in the community. One school reported making tests for 180 farmers. Another had tested for the elimination of unprofitable animals in sixty herds during the year. In such cases, the work becomes a community service, but from the educative point of view becomes excessive.

In the special schools, the possession of live stock and complete equipment allows an elaboration of work in the dairy laboratory. To the testing of milk, skim milk and cream, for butter fat and solids, testing for adulteration, for acidity, is added work in pasteurization, sterilization, standardization, cooling, skimming, butter-making, and even cheese-making. In the larger schools, the laboratory work is not to be distinguished in content or method from that given undergraduates in colleges.

In Farm Management, laboratory work consists almost wholly in the working of problems in cost accounting, planning of rotations, and mapping of farms and buildings. To this is sometimes added 'survey' work on the home farm, or on neighboring farms. Two schools had completed a community agricultural 'survey'. Again, field surveying, with compass, transit, level or plane table may be a part of the work, though this is infrequent. The class may recommend systems of cropping or rearrangement for farmers, thus performing a community service, but is not often called upon to do so. Most of the problems are fictitious, but a tendency to make them real is apparent.

METHOD IN SHOP WORK. *See Table 30*

Thirteen schools in the high school group did not include shop work in the course of agriculture, and the preparatory school likewise. All other schools gave shop work.

In eight high schools, three state-aided, the formal tool process sequence of the old-line manual training room prevailed, as in two of the special schools. In seven high schools, three state-aided, and six special schools, procedure is by process sequence until such time as the student is considered trained sufficiently to undertake a project of his own. The period is variable, some instructors allowing the student to choose and own his own project early, after a relatively brief preliminary exercise in the use of tools; more withholding freedom of choice till a relatively prolonged experience in the use of tools shall 'justify' the undertaking of elective tasks.

The thirteen remaining schools, all but two organized under the home project scheme, made undertakings in shop work either (in one case) wholly a matter for the student's choice, or (in twelve cases) optional within the range of relation to the home project, school project, or class work in agriculture. In these the student begins at once to make something for which he sees a definite need. For instance, the first shop work of a boy who has bought pullets for a home project may be to build a poultry house; if he has a house, to build appliances for use in it. The first work, in two instances, was to build the bench at which he is to work.

All sorts of undertakings were viewed in the shop work. Favorite were poultry houses, feed hoppers and troughs, brooder houses, cold frames. But bread boards, necktie racks, towel rollers, clothes hangers, fancy clothes poles, were, perhaps, more frequent still. Making products for sale or for gifts to the family were common. In a Congressional district school, the majority of boys were engaged in making kitchen cabinets, of excellent workmanship, for their mothers. Already they had furnished the bureaus, tables, book-cases and chairs for the principal's rooms. In a county agricultural school, the favorite project was a cedar chest, the market price of which might run to thirty dollars. In a state school, preceding classes having equipped the farm with wagons, sleds and wheel-barrows, present classes were turning them out for sale.

School needs were sometimes the basis of shop work. One class in a high school was completing the equipment of the building in individual lockers, another was building benches and cabinets for a chemical laboratory. In repair and construction work about buildings and on the farm, the range of undertakings was very wide indeed.

Rule of thumb procedure, the following of explicit directions in the shop work, is usual. Granted the choice of a subject, the student must undertake its construction according to a set formula. In the process method step one, step two, is almost invariable. As to results, generalizations are difficult. As a whole, the shop classes were among the most interested groups visited; in product, the workmanship of the special schools was superior to that of the run of high schools. Superior equipment, special teachers and, it may be, longer drill in the use of tools, contributed to this result. As to the superior ability of these more skilful workmen to meet original problems, no statement can be made. For the project workers of the high school, it may be said that they were solving original problems in fairly adequate fashion. Numerous examples of very crude workmanship might be cited, but no unusable product appeared. As the boys carried home their completed manufacturings under the project method, the range of exhibits was very much less than in the more formal classes.

School Projects and the Use of Land

Activities more or less definitely organized as so-called school projects are reported in the case of ten high schools and two special agricultural schools. Such projects call for the assumption to a greater or less degree of a common responsibility by all agricultural students or members of a class. In the coordination of cooperative effort, and in parallel to type activities of groups in farming they vary widely. In three schools a laying contest, in which each student entered one or a few fowl, made a 'school project'. By pairs students took charge of care and trap-nesting over a week or more. The only group end, perhaps, was that of securing a fair competition. In four cases, the group was in complete charge of an orchard from pruning time to harvest. In four cases, the group was in charge of a common plot of ground on which a single crop of alfalfa or several varieties of vegetables were grown. In one case, the class in horticulture had organized itself as a club for social enjoyment which derived its funds from the sale of greenhouse plants. In another, a class in farm crops carried on a wheat variety test. In another, every student was assigned to the survey of several farms to the end that the group might complete an agricultural survey of the community. In two cases, the group was responsible for the conduct of demonstration plots; in three, they were engaged in the construction of a hen-house or shop on the school grounds.

One more ambitious project is, perhaps, worth recording. The poultry class had built a hen-house and yard, and installed a flock of forty-five pullets, with money borrowed from the school board. They had collectively and individually engaged themselves to pay back the eighty-five dollars borrowed in equal annual instalments during the four years of the course with interest at five per cent.

The most significant school project, however, was the formation of a cooperative society for the grading and marketing of products raised by the members in their home projects, on the plan of the fruit growers' association among adults. The organization was at its inception, and no measure of its success is yet possible.

School land, used otherwise than as indicated, was usually divided into individual garden plots, either for all students or for boarding students, or for city boys who had no land for home projects. In the two cases when allotment of land to boarders had been made, the summer work of production was in the hands of hired labor, the expense of which was charged to the sale of crops. In several schools the use of the school plot was wholly for the provision of laboratory material, particularly varieties of legumes not common in neighboring farms. In boarding schools the use of the farm was ordinarily to provide for the boarding house, and was mainly in charge of hired labor. Demonstration for farmers is the end of the farm in several of the special schools, commercial profit in two philanthropic schools. The educative aim is absent in these latter cases though, as has been shown, the educative use may be considerable.

Two high schools reported giving up the school plot because of losses through theft. In three cases the possession of land had no significance. It was not used for any agricultural purpose. Principals reported that since most of the productive work must be carried on by hired labor in the absence of pupils over summer, the expense of cultivation was not justified.

Method in Outdoor Work. *Table 30*

As has been noted, weather and season made opportunity for observation of outdoor work in about half the schools visited unfavorable. If, in any one particular, the run of teachers were uncertain and somewhat reticent in respect to their replies, it was as to the amount and character of field work. The impression is distinct that, apart from certain home project schools, and some schools with farms, the proportion of time allotted to outdoor work is very small. All schools that reported any outside work whatever are included in the table, even though that outside work be confined to scattering observation trips to neighboring farms or industrial plants, as was certainly the case in at least twenty-five per cent. of the schools, or to a lesson or two in pruning or spraying, the planting of a garden in the spring or gathering fruit and field crop specimens in the fall, or judging a few horses or cattle in the neighborhood. One teacher reported that he spent every pleasant day of spring and fall outdoors, and that he thought fully one-half of the school time of his year was so occupied. This was the maximum. Several reported giving one day a week during fall and spring to outdoor work; more said, "I take them out whenever I can manage it," which seemed usually to be not often. It is to be noted, however, that in those high schools in which supervised home projects are required over summer, the total of time spent by students in outdoor work may be considerable. Reports of projects indicate from one to two hours a day during the growing season as common.

In the schools possessed of farms, the requirements already noted of a minimum period of farm work, ensured some participation in productive processes. The standard for such schools appears to be thirty hours per month of farm work, but the total of work is frequently less than in the home project schools.

In all high schools, outdoor work is in charge of the teacher who conducts classes in the schoolroom, and the arrangement is usual in the schools of Groups B and C. However, in three instances such work was in charge of others.

In twelve high schools, and one state school, in which the aim was distinctly acquirement of technology, outdoor work was almost wholly observation, usually with notes. In eighteen high schools observation was accompanied by, or subordinated to, practice in agricultural processes. That is, students did not only witness processes of spraying, pruning, judging, selection of seed, working of farm machines, and the like, but, usually after demonstration by the teacher, took part in such activities themselves, not often sufficiently to attain any marked proficiency, but sufficiently to give a first-hand acquaintance with the nature of the work. In a few cases the organization of pruning or spraying squads, of judging teams, and the performance of community service gave sufficient practice to the attainment of unusual skill. Two boys were reported to have earned their school expenses by testing farmers' herds during the summer, another had purchased a spraying outfit and made a business of spraying for neighbors, several carried on an intermittent business in pruning orchard and shade trees and hedges, another had passed his summer at the stock-yards as assistant to a stock buyer.

Farm work, determined by season and the needs of the farm, made up the outdoor work of six high schools and ten of the schools of Groups B and C. In small schools no grouping of students was necessary. In the larger schools students worked in groups by assignment. A more or less definite attempt at rotation in processes wrs reported in four of the special schools, in order that every student might partake at one time or another in all phases of work carried on. For instance, in one state school, the director, who made all assignments of work, kept a record of such to the end that every boy might have a part in the preparation of the corn land, planting, cultivating, cutting, filling the silo, and feeding silage and ear corn. Under such an arrangement more difficult tasks are assigned first to the most experienced.

Only three schools reported no outside work whatever.

From work witnessed, the impression was gained that, from an educative point of view, the outdoor lesson is ordinarily without organization. It lacks definite aim or plan. In one county school a class in poultry was engaged in a so-called 'school project'. Some twenty boys followed the instructor to the tool shop, gathered up hammers and nails, and repaired to the hen yards. Two or three boys were assigned to each yard, and busied themselves in renailing the base-boards of the fences, while the instructor looked on. In a state school a squad was engaged in digging potatoes. Armed with forks and hoes they marched in line across the rows, many of them chopping every time into the center of the hill, or spearing tubers on their forks. The only direction given was to 'throw out the rotten ones'. The descriptions appended suggest the same lack of instruction and heed to the essentials of the skill, however simple.

The Home Project Plan

The use of the home project as a means to teaching agriculture is growing very rapidly. Chief credit for its development and spread is due to Dr. R. W. Stimson, of Massachusetts. The scheme of teaching agriculture through the home project in every school where it was found was based in features upon plans first published by him. Nearly every teacher, by the method, had in his possession and used one or all of the bulletins published by Dr. Stimson, through the United States Bureau of Education or the Massachusetts Board of Education. Conceptions as to the place of the project in agricultural education differed decidedly from that of Dr. Stimson, but the practical working out of the plan has, in general, much that is common to the Massachusetts schools.

In brief, the home project is an undertaking by the student on his home farm or village lot of real agricultural work. The work may be one of production through care of plants and animals, of investment in the form of construction or improvement, of money and labor, or of comparative experimentation. In the vast majority of cases, the work is productive, with a frankly economic motive; frequently, as a necessary concomitant of the productive effort, investment in construction, with a motive also economic; rarely improvement with the æsthetic motive, or experiment with a view to the selection of improved means to economic production. As examples, the students may grow one or several crops, keep one or several kinds of animals, construct buildings or drains, undertake a beautification of the home grounds, a betterment of sanitary conditions, or try out schemes of manuring, or compare varieties or strains of plants on the basis of economic returns. The project may be seasonal, all the year round, or continued through a series of years, and the student may carry on several projects of any or all kinds, according to the means and the time at his disposal.

The range of projects is suggested by the appended classification of 1,253 students' projects. The classification for 862 is furnished through courtesy of Mr. L. S. Hawkins of New York, in charge of Agricultural Education for the University of the State of New York, the classification of the remaining 391 is compiled from lists furnished by Dr. Stimson for Massachusetts schools, and gathered by visits at some of those schools.

The prevalence of the vegetable garden project in Massachusetts and of the poultry project in New York is a result of the dominance of the state requirements for state aid to high schools for vocational agriculture. In the one case the state outline calls for vegetable gardening in the first year, in the other for poultry raising. A large proportion of students in both states are still in the first year of work in agriculture.

The carrying of home projects through the growing season or throughout the year may be required of the student either with or without supervision by the teacher, or may be optional. Fifteen high schools and two special schools among those visited required projects of all students through the course for the attainment of credit, seven high schools made the home project a matter of election with the student. In all cases, optional projects were unsupervised, in all but three cases required projects were supervised. Five schools had laid definite plans for the adoption of the project method.

In theory the project is regarded in one development as the central activity of the pupil, out of which arise necessities for adjustment calling for skill and knowledge which must be acquired, and giving motive for research, discussion, and classroom work to the end of resolving difficulties in achievement of the main end. In other words, the project becomes a problem through which the student is led into knowledge by which he may solve both it and other problems in the agricultural field. The center and core of teaching is the project. Selection of experience is radiate from the source.

In another view, more widely prevalent, the project is born of classroom, field and laboratory teaching. The student undertakes a project only after he has knowledge sufficient to intelligent choice of an undertaking and to success in an organization of means to its completion. The project is emergent rather than central, the fruit rather than the core, a verification or exercise rather than a problem.

Now, in practice, as revealed in the schools studied, the attitude of the state department or the officers in charge of the organization of agricultural education, seems to make very little difference in the use of the project. In Massachusetts schools, where the state department stands for the central problem idea, it is customary to undertake the work of vegetable gardening, as nature dictates, in the spring of the first year, after the student has been studying agriculture for some months. It is not necessary, even, that he choose his particular sub-project in gardening until he has been studying for some time. During the fall, he studies vegetable gardening 'projects' on paper in the form of typewritten question outlines. Thus the real project is emergent from previous study as is proper, according to the point of view taken by the authorities of New York and other states.

In recognition of the fact that classroom study, even under a topic question organization, tends to become academic to a degree in the absence of a real problem, two teachers had introduced sub-projects with animals beginning in the fall, in order to have somewhere in the boy's school experience a first-hand problem that should give motive and vitality to school work.

On the other hand, the proposed initiation of the study of agriculture with a home project in poultry, as in New York, makes for the possibility of an immediate problem. To be sure, the boys are expected to study poultry first, and to keep poultry later, so that the spring is the normal season for the emergence of reality. But in four New York schools, teachers had seized upon the opportunity for beginning with the realities by getting their poultry projects under way in the fall. Topical organization, much like that of the Massachusetts schools, except for the omission of the question mark, appeared in some of these schools, but direct dependence upon the text as a guide, at least to the sequence of topics, was usual. Yet two of these men had come, somehow or other, to the idea that the real problem should control. For instance, both, having poultry already purchased by pupils, and, at least temporarily housed, had yielded to the pressure of common sense rather than to the 'logic of the subject', and begun at once the study of feeding. Even in schools where the living creatures were not yet in possession of students, the most common of shop-projects, begun in the fall, were poultry appliances, and the construction or improvement of poultry houses. Fortunately for the spring arrangement the usual topical order permits a construction problem to emerge prematurely.

There is a suggestion of the prevalent academic quality of the topic or question outline method, in these cases: At one school, for some years renowned for its project work, the teacher laid upon the table many sheaves of project outlines dealing with various crops. He told the boys to look these over carefully with a view to choosing from among them those best adapted to their home conditions or their personal likings. For a half hour, while the teacher was absent, the boys sorted and selected. At the beginning there was a scuffle to secure outlines from the pile. The eagerness was inspiring until the cause was revealed. To the successful in the scuffle, possession of a thin sheaf gave delight. All inordinately thick sheaves were sorted out and laid one side without opening. Then perusal began of the thinner sheaves. Among these, choice seemed largely determined by the number of references included. The fewer, the more pleasing.

In another school, where the teacher showed an unusual independence of recommended organization and gave his approval to projects only in accordance with their fitness for the particular student and his home needs and conditions, one boy was hard at work upon a sheep-feeding 'project'. Inquiry revealed that he was a village boy, having only a small yard for project use, that he knew of no sheep in the vicinity, though he had seen some at a fair in the fall. It was his hope that, when he graduated, two years hence, he might go to his grandfather's farm in another state and become a sheep farmer.

In another school the teacher confessed that he found it necessary always to review the 'projects' studied in the fall before the planting time. Otherwise, and even then, he said, mistakes as to width of row and depth of planting were common. "We teach eighteen inches in the fall, but they are quite likely to plant at twelve inches in the spring, if we don't look out." There is food for reflection in this comment, not only with regard to remoteness of outdoor and indoor work in time, but as to the selection of subject matter for 'preliminary study'. Rule of thumb directions lack through content, as rote memorization tends to lapse, and in this form, under modern conditions, are frequently unnecessary. To drill in the formula for Bordeaux mixture, for instance, as is frequently done, is absurd.

So far as classroom work was observed it did not appear that the formal question outline development of the project plan was in anywise superior to the categorical topic outline with its references. Nor is there any particular reason why it should be. A reading of many such outlines leaves the impression of a trend to a stereotyped form characteristic of the 'subject' rather than of the particular problem. Indeed, several such outlines studied are palpable question developments of the text-book treatment. Some outlines, however, recently published by the States Relations Service show decided improvement in both form and development. A mere change from the categorical to the hypothetical form of statement does not of necessity secure interest or provide motive. Teachers who made their own topic assignments seemed to secure quite as good a response as those who used the mimeographed 'projects'.

But the topical assignment, under either method, possesses one advantage over the common text recitation. It brings the student into contact with a variety of sources. Different presentations, attitudes, points of view, deductions and inductions, come to his notice, and the real use of the book as a means to the solution of problems may come home to him. Unfortunately, however, the intent to teach the use of books, is not very evident in the usual development. References are, for the most part, specific, by title, author, chapter and page. For the farmer nobody stands by to furnish such direction when a new question arises. If books are to help him, he must know how to find the material he needs. One teacher reported specific recognition of this fact. He said, "For the first half-year I give the boys exact references, and I make it a point to show them how I found those references to give them. After that, I expect them to find their own references, when I assign a topic, and they do. They can use the card catalogue and the index as well as I can."

The impression is distinct that academic treatment and lack of correlation between indoor and outdoor work is still marked, even in home project schools, whatever the expressed attitude of authority with respect to the place and function of the project. But it may be that the conception of place and function in the mind of the teacher will tend to lessen these evils. Indeed, in the cases cited, such appeared to be the case. Accordingly, which point of view the teacher accepts becomes a matter of importance.

It is very plausible to say that the boy cannot choose or carry out a project until he knows how. That is exactly the attitude of the shop teacher who holds that the student must not be permitted to make a feed hopper until he has mastered the tool processes. It tends to make the project an end rather than a means, and such it is plainly in the minds of many teachers and pupils. Certainly the motive to study lettuce growing this fall in order to make money next spring furnishes a more immediate motive than the vague or supposed desire to become a successful farmer. But it lacks the immediacy of motive of the problem that demands an answer *now*. As has been suggested, the process sequence may result in a more finished product, in wood or iron, than the problem method. But if the self-adjusting human organism be considered the product, such superiority is by no means clear.

Further, the statement that the project should be the emerging resultant of class teaching carries in itself a contradiction. The project has been brought into use in the belief that class teaching does not teach. But making it an end is to presuppose that class-room teaching does teach. Certainly, then, the project becomes, at best, a matter of supererogation, an exercise or confirmation.

Now the introduction of 'doing' is not for the purpose of giving skill, but of giving meanings. Not one teacher in the lot professed an expectation of turning out skilful farmers. Every one expressed the hope of turning out intelligent farmers. That being the case, an apperceptive basis for learning, founded in the vicarious experience of the classroom is, at best, but poor preparation for the conduct of a successful project. On the other hand, the first-hand experience even of an unsuccessful project may be made of first-class importance in giving meaning to classroom teaching. It is a common complaint of teachers that "I can teach the farm boys all right, but the city boys and the girls give me a lot of trouble." Surely. Because the words of the text and the classroom convey a meaning of some sort to the farm boy, if little or none to the city boy or girl. Thus, the project is a means to the perfection of teaching, not an end to be attained by teaching. To make it the fruit rather than the seed or core of method, is to discharge it of its chief function.

In some schools, particularly in the Middle West, this argument against the project is made: "This is not a region of specialties. Our farmers are general farmers, they farm on a large scale, and a diversified plan. They are not growers of this or that, but farm managers. Besides, they are shifters. It does not follow that because a boy goes to school in this county that he is going to take up a farm even in this state. He may go to Canada or the Pacific Coast or to the Southwest. If you train him for a specialty here you waste his time. All we can do is to give him the principles that guide in successful farming. We cannot give him a farm to manage. The home project is all right for the East, but it is no good here."

But this is to say that because we cannot give a boy complete experience in the problems of life, we should not undertake to give him any. The project, properly used, is a means, through typical instances, of induction into the principles that govern successful farming. In so far as it is real and first-hand it is superior as a means to the vicarious experiences of the recitation and lecture and observation trip. The spread of the home project to such states as Indiana, Wisconsin, Michigan, and Minnesota shows a growing realization of the ineptitude of the contention.

Nevertheless, the charge that the home project is an isolated experience that does not carry over principles into general application is not without foundation, as the following instance may illustrate. In a school where the adaptation of projects revealed unusual skill, a fourteen-year-old Italian boy had purchased three cows, to be kept under housing conditions that would appall the ordinary dairyman. By intelligent feeding he had increased production to a profitable basis at the ordinary price; by extraordinary precaution as to the cleanliness of cattle, lintel, and utensils, he had produced milk of the highest grade. Board of health requirements hold to a maximum of 10,000 bacteria to the cubic centimeter. Tony showed the visitor certificates for samples of his milk showing less than 5,000. By virtue of them he had raised his price to twelve cents a quart. It is plain that he had been taught to do one thing well, and he was enthusiastic over the possibilities of farming.

Now, in a yard adjoining the barn, ran half a dozen hogs. Among them was one particularly scraggly and stunted. In reply to the question, "What is the matter with that hog?" Tony said, "I don't know. He doesn't seem to grow. He is two years old already. We are keeping him till he gets big enough to kill." A primary principle of the economics of animal production should have told him that every meal that hog ate represented a dead loss. But the knowledge that guided him in the selection and management of his cows was of non-effect with respect to the hogs.

The fault lay not in the project, but in the use of it. Elements common to both situations existed. The principle involved in the selection of profitable cows is not different from that in the selection of hogs. Moreover, it was concretely illustrated in his own experience so that exposition of it in the classroom should have carried to him more than a casual recognition. The case but serves to emphasize the need for pedagogical skill among teachers of agriculture. It is perfectly possible, in theory, for a teacher perfect in knowledge and skill, to lead, from a single problem, his pupil through the whole realm of knowledge. But there are no such teachers. For regeneration of the school, a first requisite is to provide teachers of knowledge and skill sufficient to the selection and emphasis of essential principles. Such teachers, working through the concrete experience of the pupil, as obtained in the home project, have decided advantage of those who cannot know the pupil's experience.

As has been suggested, the selection of projects appears largely a matter of dominance by central authority. The justification of a scheme which calls for a uniform type of project in a given year is made upon the grounds: *One,* In order that there may be a basis for classroom recitation there must be study of topics common to the whole class. *Two,* There are activities common to all farms which are within the scope of abilities and means of students, *i. e.,* the farm garden and the farm flock. Hence, the prevailing dominance of the kitchen garden and the poultry project in state plans.

The first assumes that the recitation is necessary. The second may involve waste and error. Now the classroom teaching may, and should, prove an important and, perhaps, the most important means of organizing the material of individual experience. In so far as the experiences of pupils are identical, the simplification of the task is real. In so far as the common experiences of the class are limited in type and in number, the trend to a deductive treatment is likely. In so far as common experiences are numerous and still more as individual experiences are diversified, the opportunity for inductive teaching comes. Thus, enlarging the scope of the uniform project increases the number of instances involving common principles. Increasing the diversity of individual projects is, obviously, more easy of accomplishment than enlargement of the scope of common projects.

The selection of a specific situation is useful largely as that situation contains elements common to probable situations in future to be met by the pupil. The project is, or should be, a typical instance. But the single instance may well fail, and commonly does fail, to reveal its class bearing. A multiplication of instances involving a common element compounded with differing concomitants, serves to the exclusion and clarification of the common typifiers. If the single project has a weakness as a means to teaching, that weakness lies in its singleness. To induce a central principle diversity of instance is greatly helpful. Thus, as a help to inductive teaching, the multiplication of experiences among members of a class of pupils, though it involves experiences vicarious to an individual, serves concretely to ease the process.

The project method is essentially inductive. To reduce class experience to a single or small group of instances, and to place those instances under an abstraction voiced by the teacher or text is the easy way for the teacher. To induce the generalization from apparently diverse instances is to increase his difficulty. But the school is not set up for the ease of the teacher.

The skill and knowledge necessary to pick the essentials common to apparently diverse agricultural practices must be of a high, but not necessarily of a superlative order. Principles of control are, after all, relatively few, if multifarious in their manifestations. The educing and organization of these principles from as wide a range of real activities as possible is the function of class teaching in agriculture. To subordinate and lessen the number and scope of these activities to meet the needs of 'the subject' is to prostitute that function.

Now the uniform requirement may not only fail of providing adequate adjustment to the needs and means of the individual, but may result in maladjustment. A boy resident on a large dairy farm, expressing a prime interest in cattle, desirous of succeeding his father in the business, is confronted with the necessity of carrying out his initial project in poultry. Partly in pique, partly in curiosity he chooses with approval, the care of ten pullets under a famous advertised scheme designed for the square rod of the suburbanite's backyard.

In another district, the sons of an onion grower carry projects in kitchen gardening. During the growing season every member of the family is engaged all day in the onion fields. Only before breakfast or after supper can the boys be spared for work in the garden. Further wheelhoeing and hand weeding they find uninstructive and uninspiring. The teacher has taken the bit in his teeth and declares that such forcing shall not occur again. The boy on the onion farm shall have an onion project if he wishes, the boy on the dairy farm, a dairy project if he so desires, and the son of a fruit grower an orchard project.

In another school where the full four years of work are under way, only market garden, poultry, and fruit projects are in hand. The community is one of specialties. Complaints of pupils and parents against a requirement not in line with their immediate needs and interests have led to the rejection of the state outline in part, under a nominal acceptance, and selection both in project and classroom work is based on local demand.

Selection of home projects is made usually, in conference of teacher and boy, in which the parent is included to the extent that he has promised land, animals, materials, and time for the boy's accomplishment of it. Sometimes boy, parent, and teacher meet at the boy's home for a choice of projects within the limits of the established plan of the state or school.

Some teachers were exercising an admirable independence in judgment in order to adapt the nature and scope of the home project to home conditions and needs. But too often the selection to which the approval was given conformed more nearly to the demands of an outside authority than to the conditions of the case.

Some of the state requirements show an excellent judgment and a measure of flexibility. For instance, the first year of the New York plan calls for projects in poultry and shop work. Both are well within the means of the ordinary farm or village boy, both involve common farm activities, both involve active 'doing', both may give 'quick results', both offer a rather wide range of selection not closely restricted by seasonal or soil limitations. Both are based on a psychological foundation of appeal to adolescent boys. The two correlate excellently. Yet it is doubtful if the imposition even of such a requirement upon all alike is to be justified. Variation in means, interests, needs, is too great. These conditions, the teacher who knows boy, parent, and home conditions, should be allowed to weigh as determinants of choice without dictation from authority. Several teachers have reported that it was their custom to ignore authority in the matter of selection and to get projects under way independently. When the project proved successful it was in no case rejected, though consultation in advance proved quite likely to result in a forbidding of the undertaking.

The conditions that enter to determine the choice of a project are numerous and call for careful judgment. It is easier to fall back upon the outline. That teachers do make their own adaptations in some cases is a very promising sign. Local market demands, home needs for consumption, soil, climate, topography, season, capital investment, availability of tools, machines, horse power, continuity and intermittence of labor, prompt results, margin of probable success, immediacy of motive, personal preference, attitude of parent, and of the community, relative educational values, specific aim of the teacher, are some of the factors that must be considered. Except for educational values, it is not possible to say which shall control in every case. Instances of limitation might be cited at great length, but a few will serve. One boy was the son of a tobacco grower. He wished to grow a plot of tobacco. But community attitude toward the teaching of tobacco growing was unfavorable. He had to grow onions. Another boy wished to undertake a dairy project, but his father would not spare him the cows. Another wished to keep hens, but his mother disliked them. A father wished his boy to undertake feeding of dairy cattle, the boy preferred to grow strawberries. Although in Massachusetts and New York hay is the dominant crop, only four projects with hay, all experimental, are reported, among 1300 odd from those states. Intermittence of labor, and lack of scope, outweigh local demand.

Intelligently selected and used, the home project method is one of the most hopeful features of agricultural education. On the social side it moves to the unity of school and home education, it lessens the need for special away-from-home institutions, it enables a greater number to couple the interests of vocational education with interests shared by their fellows in school life, it opens a way to release from traditional school methods. That it has not yet realized its possibilities, is plain, but not altogether discouraging. A first means to that realization is the teacher imbued with a philosophy that carries beyond the subject to the pupil, a teacher who sees in the achievements of his pupils the progress of method rather than the fulfilment of ends. Through the freedom of this teacher is to come the progressive organization of the agricultural course. The teacher and the freedom are rare, both are possible.

The opportunity of the teacher in agriculture, under the home project scheme, for individual teaching is unexcelled. As no other teacher can, he comes to know his pupils. He sees the boy in the school and in the home environment. He knows the conditions under which the boy has experienced life, his personal traits, his interests, the limitation of his means, his needs, his opportunities, as one who meets him only as one of a class, can seldom know him. To select, then, according to the needs, and to teach in terms of a first-hand experience of the boy, are to the teacher of agriculture difficulties relatively easy of solution.

Supervision of Projects. *Table 30*

When supervision is required of the teacher, the usual plan is to visit the homes frequently in the summer, less frequently in the spring and fall, while school is going on, and occasionally during the winter months. Once a week in the summer, once in two weeks in the fall and spring, once a month in the winter is a common plan. But few teachers reported a strict adherence to the plan. Distance to be covered, means of transportation, weather, and other duties tended to the modification of any set plan. Often the route of the teacher is a long one. One man reported it necessary to cover 125 miles in order to complete his round of visits. As a sample of how the work may be carried out, a report furnished by Mr. A. W.

Doolittle, of the Concord, Massachusetts, high school, is given on page 182. The various means of transportation are indicated by data furnished by Mr. L. S. Hawkins for teachers in New York on page 180. The time and the expense involved in visiting projects limits the size of classes that a teacher can handle. In Massachusetts a limit of twenty pupils to the teacher has been set.

Records

In all cases where the home project was used in the teaching of agriculture, full and more or less systematic records were required of students. In some schools the student kept a daily record of his activities, even those not concerned with the project, and of the weather. Instructors said that this requirement proved irksome. The boy was quite likely to let the record go for a week or more and then to fill it out by guesswork. Even the reports of hours of labor upon the project, particularly in the unsupervised projects, were subject to the same sort of treatment. But, on the whole, the requirement of records of expenditures in money, man-labor, horse-labor and fertilizer, and the like were faithfully kept and checked against the sale of the product or the market price of products turned in for family use. A system of cost accounting is integrated with the production project, a very useful feature. Out of this, however, seems to have arisen a tendency to measure the success of the project in terms of the profit made by the student, and to look upon those schools as most successful in the teaching of agriculture, whose students have conducted most profitable undertakings. Certainly such a standard is in line with that of common life, wherein the most successful farmer is he who makes most money. It tends to the conviction of many parents of the worth of agricultural schooling. But from the point of view of the educator, it is not altogether fortunate. It confirms the view that the project is an end rather than a means. It measures success in terms of the material product rather than in terms of education of the boy. When this view is held, it is almost inevitable that the project be reserved as a test of ability in the use of skill and knowledge, rather than that it be put forward as an immediate means to the acquirement of skill and knowledge.

With the question and topic outlines of project plans the use of the note-book is common. After reading his references the student is expected to organize in his own words the procedure that he proposes to use, that he is using, or that he has used. Of the value of such a plan there is little question. But sometimes the note-book is overworked. In one school the plan called for three note-books. In the first, the student answered all questions out of his own unaided experience. In the second, he entered answers derived from reference study and inquiry. In the third, he organized the whole in his own words. One student had filled fifteen five-cent note-books in the course of a half year.

EXTENSION WORK

No study of the extension activities of teachers was made. As a rule, the interest of teachers in this side of their work is strong. A favorable acquaintance among farmers may lead to the county agent's job, which pays a higher salary than is usual for teachers, and such acquaintance is distinctly an aid to the agricultural course, both in attracting sons of farmers to it, and in securing co-operative interest from farm owners. It may lead to subordination of the teacher's interest in teaching to that in community work. Several teachers objected to required supervision of projects on the ground that it interfered with work among the farmers. Others held that the project was a most useful instrument to extension teaching. Through interest in his son they reached the farmer, by more frequent visits they became well acquainted with him and his problems, and were able to give and to receive more aid among the few thus reached than by a more cursory acquaintance with a greater number.

In itself the extension work presents a large educational problem, that is beyond the scope of this paper. It is to be noted, however, that it is not of necessity educative. One teacher reported with pride that he had repaired for the farmers of his county eighty-nine binders during the year. Whenever a farmer's binder broke down, he telephoned to the teacher to come out and mend it. The inference is obvious. In Minnesota the instructor is required to give one-fourth or more of his time to extension, projects are likely to be unsupervised. In Massachusetts no extension requirement is set. Supervision of required projects is made law.

CORRELATION. *See Table 31*

Attempts at making common bonds between the divisions of productive agriculture and other subjects in the course for students in agriculture are reported in only half the schools. Where agricultural 'bias' was said to be given to any subject or where it was given a distinct appellation, as 'Agricultural Botany', the fact is recorded in the table. Botany, zoology, arithmetic, chemistry, and English are the most frequently noted. Relatively the adaptation is much more frequent in the special schools than in the high schools. The modification of English, however, is commonly restricted to the short courses, since modification in the regular course is fatal to high school credit.

The degree to which modification takes place is uncertain. In lessons observed the reported 'bias' was frequently imperceptible. As an instance one high school teacher who reported enthusiastically, "Yes, I tie up all my work with the agriculture," gave a lesson on the endogenous stem, at which the visitor was present. The day was hot and the windows open. Just out of reach the grass grew a foot in height. Across the road some hundred feet distant the corn stood in green rows. But the center of the study was a diagram of the stem of the palm tree. Another made his lesson in Farm Arithmetic a drill in solving problems all of one type, like this: "What number diminished by seventeen per cent. of itself is 166?" Another, who reported his chemistry as close to agriculture, was drilling his class in the writing of equations such as $2\ NaCL + H_2SO_4 = Na_2SO_4 + 2\ H\ CL$. Except as the chosen text carries an agricultural selection, differentiation in teaching is not marked.

On the other hand, one teacher, who apologized for seeming to neglect the pure science aspect of chemistry in a particular lesson, excused his emphasis upon the salts of ammonia and nitric acid on the ground that the boys might find a knowledge of them useful. Some suggestions of the closer correlations achieved has already been noted under Course of Study.

Correlation between work in laboratory and that in recitation or lecture is ordinarily very remote in time, and frequently in substance. A class which spends half a period in recitation upon the capillary water of the soil may give half the term to soils experiments in the matter, and the present work of the laboratory and the classroom be weeks or months apart in relation to one another. The following of set outlines of experiments in order is responsible for the remoteness in time, the selection of experiments with regard to convenience of demonstration rather than importance of the matter demonstrated is probably responsible for much of the remoteness in subject matter. Definite selection, however, is made, that the work of classroom and laboratory may bear a meaningful relation to one another in fourteen schools. The method of accomplishment is usually demonstration by the teachers (compare Table 29).

The subject matter of class study and the outdoor work seem to have no necessary correlation with one another. The one is determined by the text, the other by seasonal convenience or farm demand. However, in six schools definite attempt is made to bring outdoor work and classroom study into close relation, both in time and in substance. All but one of these receive state aid for agriculture on the home project basis. Eight schools, all working on the home project plan, succeed in bringing about a close relation in content between class study and field work, though the two remain far apart in time. Such schools study the details of the 'project' at one season, perform the work at another. A class studies all the details of lettuce growing in October, and plants its lettuce in April or May.

So long as the text dominates in the classroom while the realities of season and location dominate outdoors the remoteness of these two factors in the teaching of meanings will remain. No more marked weakness in agricultural teaching exists. And the inclusion of project schools in the group undoubtedly makes the showing here better than the normal.

ADJUSTMENTS

To Seasonal Requirements

In analysis of the organization of the course of study the frequency of adaptation to season, in particular of the crop subjects has been noted. In the outdoor work a more common adaptation may be expected, and is found. In Table 32 it will be noted that all but five of the fifty schools have made a definite selection of outdoor work in accordance with seasonal requirements. In seven high schools and one special school the dominance of the 'subject' has been set aside, and the outdoor work planned by season. For instance, in the fall all outdoor work is concerned with harvesting, hill selection, or the study of varieties in the field.

Adjustments to Local Demand

A study of the productive subjects, in so far as it is complete with reference to the content of 'the subject' agriculture, almost of necessity touches in some phase upon those features of local farming that are important, but that accidental impingement may not result in emphasis being given to such phases. Dominance by the subject in this fashion has been designated by inclusion under the caption *Slight* in Table 32.

Yet where the dominance of the 'subject' is on the whole very marked, the intelligence of instructors may result in emphasis upon the points of particular importance in local farming. Twenty-two high schools and four special schools do make such emphasis in particular divisions according to the character of local farming. Dairying, where such is an important local industry, receives special emphasis in fourteen high schools and three special schools, Fruit Growing in ten high schools and one special school, Poultry in four high schools, Gardening in one high school, Forestry in one, Road Building in one, Drainage in one, Corn Growing in two high schools and one special school.

When the 'agriculture' represents a definite organization with respect to the characteristics of local production, schools have been grouped under the caption *Marked*. Six high schools and two special schools have been included. In two of the high schools selection of topics has been made wholly with regard to what, after preliminary survey, have been determined as local interests. Not only are divisions of subject matter of local interest emphasized, but such as are not of immediate local interest are omitted from the program.

In this table it is worth noting that local adaptation is a feature of schools state aided, most of which are organized on the project basis; that it is not a feature of the unaided high schools, or of the state schools, with their large units of community service. Indeed, a swapping of courses among the state schools would not materially interfere with their usefulness. With two exceptions, a change of location among the special schools would call for no material changes. Their course is based upon the needs of 'the subject', and, as has been noted, is an approximation to the agricultural college course, in respect, at least, to the productive side.

Other Adjustments

Particular adjustments of course and method have been noted throughout the study. The hobby or specialty of the teacher may effect selection and emphasis. This was a marked feature in two cases. In one, the teacher, whose only adequate knowledge lay in the field of market gardening, imposed the study of that topic almost to the exclusion of others, in a remote dairy region. In another, the teacher began his course with the study of Farm Management.

Definite selection and emphasis upon topics chosen because of their adaptations to the needs of local progress appeared in two schools. In both cases dairying was a dying business in the locality, though general; fruit growing a promising business, though rare.

In at least three schools the principal effort of the instructor seemed to lie in the direction of producing materials for exhibition and advertisement. In each case the camera was an important adjunct to the agricultural equipment.

One city high school requires of its students three months of farm labor some time during the four years in order to graduate.

A town high school in which about half the agricultural students enter after Thanksgiving and leave early in the spring, has devised a scheme of quarterly alternation of topics, so that such students may receive complete subjects. Students entering in September last fall studied corn during the first quarter, late entrants joined them in the study of small grains during the second quarter. Next fall September entrants will spend the first quarter on small grains, late entrants will join them in the study of corn in the second quarter. Thus, without duplicating classes, the short term boy covers the subject of Farm Crops in two years, the full term boy in one year. Recurrent treatment is not common, but adjustment to that need is found in the plan of the course of study in several of the special schools. Diversification and intensification of special topic study follow the general elementary treatment of the first year.

The director of a philanthropic school, in which farm work by assignment constitutes the major part of agricultural training, had noted that the daily tasks lacked the inspiration of direct motive and tended to a tedium far from educative. As a means of overcoming this difficulty, he proposed to make it incumbent upon each teacher to call together his class before the beginning of daily work and to explain to them the purpose of the task, the opportunities for acquirement of skill and knowledge in it, and its meaning with relation to the management of the farm as a whole. The plan was not yet in effect, but the making of it is significant as recognition of the need for motive and organization in outdoor work.

DISCUSSION OF GENERAL APPLICATIONS

The study has given some glimpses of the ways in which schools have sought to adjust their means to the ends of agricultural education. What shall be taught and how it shall be taught are the larger questions that they have sought to answer. In general, they may be said to be in some measure of agreement with regard to answers to both questions, particularly the first. That which shall be taught is the more or less formally organized content of science with respect to production handed down by the agricultural colleges, along with the more fully standardized content of academic subjects selected by the high schools under college dominance. That content shall be taught, in part, at least, through the use of text-books and lectures, and the question and answer recitation method of the high school. Yet the variation in attempts at organization and modification of method in the contact of teacher and pupil reveal the fact that the solution is by no means 'standardized' for the agricultural course, and that active intelligence on the part of teachers and supervising bodies is at work to the determination of closer adjustments to the needs of pupil and community.

The factors that make for determination of content, organization, and method in the secondary course in agriculture may be classed as social factors, psychological factors, and miscellaneous factors of expediency due to physical and administrative situations in large part. But the classification is not easily made satisfactory. The terms used are not wholly disparate, and the factors distribute themselves under one or more heads in nearly all cases. But the attempt is made for the sake of organic treatment to consider some of the factors resultant in the agricultural course, under that arbitrary system.

The aim of the agricultural course has been seen to be variant, with a predominance of the vocational aim. In the state-aided and special schools in particular, education for farming rather than education for the farmer seems to be the prevailing concept. The economic aspects of rural life rather than the diversified activities of rural society as unified in the farmer have served to the determination of aim. As has been noted in previous discussion this fact is not wholly fortunate. Yet to this fact is due, probably, in no small measure, the success of the agricultural course. The content of the so-called science of agriculture has to do at present almost wholly with production. Thus, for the vocational aim, an organized selection of productive principles is at hand. Yet, even in this organized body of knowledge selection toward the end of control of production is not yet closely enough adaptive to the needs of vocational agriculture in the secondary school. Much that is irrelevant or remote to control is included in the content of productive agricultural subjects. If the accepted content is to be made of most use in respect to its aim a closer study of relative values is necessary, and a rejection of superfluous material by the criterion of utility in control over the processes of production.

If the aim be to enable the student to gain a knowledge of farm life sufficient to intelligent choice of farming as an occupation, then the scale of values becomes different. No longer is it a first necessity to give that knowledge which leads to profitable control of plant and animal life. Rather the appreciative aspect, both emotional and intellectual, becomes dominant. What it is that makes the life of a farmer worth the living rather than what makes that life possible should be taught. But little hint of the difference appears in the schools of the study. The same content, somewhat abbreviated and somewhat more academic, is found under the prevocational aim.

If the aim be preparation for the life of the farmer, then the curriculum as a whole must be subjected to evaluation in the light of the needs of the individual and society in a rural environment. It is assumed that provision is made under this evaluation for the farm boy by the inclusion of those subjects in his course which the city boy studies. After all, says the traditionalist, the lad is a human like the city boy. If his vocation be provided for, in other respects his needs are the same. We can pass without consideration the question whether or no the organization of the academic work be suited to the needs of the city boy. It does not follow that it is adapted to the life of the farmer outside his vocation. He lives in an environment in many respects widely at variance from that of the city dweller, both in its physical and its social aspects. More study, such as is now being made by students of rural sociology, is necessary to the determination of social values in the curricula of the secondary school for country boys. In the light of such knowledge, a new set of values should come to determine the selection and organization of subject matter in the course. But again the aim is apparently ineffective to the determination of content. The same agriculture is taught whether the aim be vocational, prevocational, preparation for country life, liberalizing, or 'cultural'.

Even in the productive aspects of agricultural study much is to be found that leads to enlarged appreciation and enjoyment of the activities of life both in business and in leisure. Selection that ignores these values leaves unfilled a large gap in the scheme of education of a future country man. He is a man and socius even in the acts of production. Not all of 'culture' is to be lugged in by its academic ears.

But on the productive side, more immediately, selection is largely determinate in the light of local conditions of farming, present and prospective. That farming which exists in the community possesses at least a 'survival value'. It is a solution of evolving problems 'standardized' in the present. As such it has received and should receive the attention of the course maker. On the social side it may determine the probable situation of the emergent student. If not, it may be justified pedagogically as a factor in selection. Whatever the future seat of activity of the pupil may be, his present home and community situation furnishes the concrete means to learning of the principles that shall control in his future activity wherever it may be. Further the present adjustment of farming to local conditions is the basis upon which a new adjustment must be founded. The line of progress is most nearly indicated by the trend of local variation. That educator who would set aside, in the prescience of his own training, the facts of local adjustment is incautious. Selection in the light of progress must not be *a priori* but *a posteriori*. Better farming tomorrow shall come out of the farming of today.

As to what proportion of 'subjects' in the course for the farm boy should bear upon production, any arbitrary standard is likely to prove futile. It is probable that in the case of the prevocational aim production will play a smaller part in the course than where the aim is vocational. The allotment of one-half the course to productive subjects does not seem excessive in the vocational schools, though no measure of results is available for determination. The teacher and the pupil are determining factors at present. A bright farm boy may learn more of the science of production in a single unit than a city boy in twice the allotment, and a poor teacher may make the work all vain. Division into subjects is unreal and arbitrary, allotment of specified limitations to subjects still more so. Only experience can give us a norm. At present, a minimum of four units in vocational agriculture seems to prevail, with a trend to increase in the allotment.

The present lack of correlation between the productive agricultural and other subjects of the course is but an indication of the futility of arbitrary division. When the pupil is taught in the terms of his own life and with reference to the satisfaction of his own needs there will be fewer worries over the disproportionate share of school time allotted to the teacher of agriculture.

On the psychological side, age, home training, and school preparation are important factors in the determination of method. It is these that give the boy his basis for learning in the secondary school. Grouping by virtue of previous school training is found in the special agricultural schools, the usual plan being to allow the boy more advanced in school rating the opportunity to proceed more rapidly to the conclusion of his course. He may devote more of his time to 'agriculture', and even proceed at a more rapid gait through his 'subjects'. With high school training of some sort, as with a greater maturity, he is more likely to be subjected to the lecture method of instruction than the younger or less prepared boy. In view of his probably greater experience the words of the lecturer may carry to him more meaning than to the boy less well founded in experience, but that the lecture method is to find its justification upon such grounds is to be doubted. However, a comparison of method between the high school and the special school group carries the implication that it is so justified in the minds of teachers.

The distinction between the city boys and the farm bred boys is one commonly recognized by teachers in conversation, if very seldom in practice. The most complete recognition of group differences is found in one of the 'year round' philanthropic schools. There the city boy must enter in March and work on the farm over summer in order to win his promotion at the end of the year; the farm bred boy may enter in October and complete his studies with the city entrant of March. Other modifications of procedure have been mentioned.

In the ordinary schools the problem is one of exceeding difficulty.

With a fixed system of class teaching, a school year limited to the months in which production of crops is at a standstill, an equipment inadequate to the provision of farm experience for the city lad, and an overburdened teacher, the distinction between groups may well seem hopeless. A provision that special schools in some places be kept open for city boys throughout the year, or the arrangement of the course for city boys to carry over summers, while that for farm boys carries over winter might solve the difficulty in special instances. In the high schools the adoption of the required project under supervision seems the most successful adjustment in sight. A mere requirement of a brief, unsupervised and unchecked period of farm labor some time during the course is hardly likely to solve the problem. But definite selection in outdoor activities, such as has been noted in assignments of farm tasks to city boys, and the specific treatment of cases under individual teaching, are well worth while. The academic method so prevalent in the classroom must result largely in waste of time for the city boy. Even to the country lad in its common abstractness it may be of little meaning.

Provision for motive is a matter of individual psychology. What serves to rouse one lad may bore another. But there are motives of wide application. The commonest of these is the motive of gain. There is nothing dishonorable in the use of this motive. Talk of appealing to the 'loftier instincts' is fatuous. 'Lofty' may be synonymous with remote and unreachable. Certainly the motive of academic agriculture is 'lofty' enough by that measure, even though it be professedly economic. To attempt to reach the boy through interest in his own welfare is not to display an apologetic weakness. Rather it is a revelation of social and pedagogic insight worthy of the best of teachers. The success of the home project is due very much to the selection of motive.

In the home project again may appear motive based on the mere love of doing, the instincts of curiosity and manipulation. But an unapplauded and isolated doing is not in itself sufficient in all cases to carry a project forward. The reported lack of success in unsupervised projects is, perhaps, as much due to lack of satisfaction in the approval and interest of another as to lack of criticism and instruction. But the value of consecutive activities not too prolonged is hardly to be doubted. The interest and even delight of certain classes in shop work and farm mechanics was in marked contrast with the attitude of the same classes in recitation.

Too little effort, however, seems to be made to appeal to the native interests of the boy in the organization of the course of study and the method of class teaching. The determination of sequence, apart from modifications due to seasonal and local demand, is ordinarily according to the matured psychology of the adult mind, that has organized the 'subject', already known, in relations conducive to a unity of grasp. That is, the sequence is according to the scientific order, proceeding from the physically simple to the physically complex, from the inorganic to the higher organic complexities. But it is questionable that there exists one organization of the body of agricultural knowledge which is the only scientific and logical organization. To proceed from the soil to the plant, from the plant to the animal, from the animal to man is certainly scientific, but it is no more scientific than an organization that begins with the relations of man to the animal, of the animal to the plant, of the plant to the soil, or than one which relates man to the plant, to the animal, to the soil. Scientific Agriculture is an organized body of knowledge dealing with the applications of pure science to the control of the energy of plants and animals for the uses of man. In terms of energy it centers about the photo-synthetic process, yet it deals with a cycle of storage and release of solar energy from end product to end product, as exemplified in CO_2 and H_2O. The cycle is complete. It has neither a beginning nor an end, like the circumference of a circle. In agriculture it centers about the needs of man. Where, then, we choose an end point is a matter of convenience or interest and not a matter of scientific necessity. From the point of view of a psychological organization the procedure of the learner and the learned are to be distinguished. The organization of retention is not necessarily that of acquirement.

Indeed it is hardly to be doubted that the organization of the 'subject' in the mind of the most hardened of professors of agriculture is not that by which he acquired the organization. Learning is through interest and activity. Interest and activity are native. The first interest is in that which concerns the learner. Recognition of the fact appears in the use of the economic motive. But it can hardly be said that agriculture becomes unscientific if learned in its relations to man. In fact it ceases to be agriculture where not studied in those relations.

But, setting aside the motive of self interest, it is normally true that the boy is more likely to find interest in the living, moving animal, than in the living, moveless plant, and in the plant than in the apparently inert soil. In a broad generalization it may be said that the psychological sequence is not far from the reverse of the so-called logical. The boy who desires pocket money may find in that desire the motive to study the dairy cow, through the dairy cow motive to the study of feeds, through feeds motive to the study of crops, through crops to the study of soil. To set the study of the soil as the necessary first prerequisite to the earning of pocket money by the sale of milk is to remove the motive as far as possible. That the scientific development of the 'subject' is such as to require remoteness is not to be conceded. It is quite as scientific to begin the study of agriculture with milk as with water, of zoology with the cockroach, as with the amoeba.

The argument is sometimes put forward that the first rule of pedagogy is to proceed from the simple to the complex, from the concrete to the abstract. From the chemical angle $CaCN_2$ is more simple than the albumen of the egg. Physically, the rock particle of the soil more concrete than the marketing of a bushel of potatoes. But, in terms of experience the statements can be reversed. Immediacy of physical contact is no warrant of intimacy of experience. The soil upon which we walk may have a less meaning in terms of experience than the nest of the young swallows the cat caught last summer. Method in agricultural teaching is not to be determined by repetition of terms, but by the realities of experience.

On the side of expediency, both organization and method are modified by such factors as equipment, the demands of high school and college 'credit', the requirements of the state and of the textbooks. Both are limited by the distribution of the teacher's time among varied duties, and the ability of the teacher.

The adoption of the home project has done much toward solution of the difficulties due to lack of equipment, but much is yet to be done in provision of adequate library facilities. Expenditures for annual increments to the stock of agricultural reference books need not be large. Probably what is now spent for set texts is nearly as great an annual sum as need be set aside for reference books under a problem method of procedure.

The use of texts is not as a rule formally required in the agricultural subjects. If it be, the text may be made a servant rather than a master, by the teacher who has the ambition to be free.

Modifications due to the incubus of text are not necessary modifications.

So long as preparation of the few is to dominate the course and method for the many, the requirements of high school and college 'credit' are factors seriously to be reckoned with. That teacher or educator who sets the education of the pupil above the requirements of tradition risks his professional life, if he makes sure his intellectual salvation. But the organizer and teacher of the agricultural course have not the excuse of the principal and teacher in the standardized high school subjects. Their work is newer, its content and its method not yet so hopelessly crystallized, its aim less vague. That the tendency exists toward standardization of the course of study for farm boys along lines made familiar by the college entrance requirements and other soporific influences is unfortunate. The working out of an organization and method correlate to an aim based on the needs of the individual in a rural community, if successfully done, may suffice to the breaking of the crust of tradition. No man in the field of education has greater opportunity for service than the teacher of agriculture.

Those in charge of the work of secondary agricultural education under grants of state aid seem generally alive to the fact that successful adjustment of school means is to come through trial and rejection by teachers in the field. However rigid the requirements they have laid down appear, they can be regarded only as tentative. No single teacher who had demonstrated the efficacy or apparent efficacy of a variation in organization or method reported rejection by central authority. The function of these officers is to aid rather than to prescribe. Most of them are keenly aware of it. They have put forward prescriptions with the idea of aiding, and they do not stand stupidly in the way of progress. That they have aided in many if not most cases, is probable. But the same cannot be said of all prescriptions. When the state manual stands as the orthodox bible of all teachers who would survive, a real obstruction blocks the way of the teacher of agriculture.

That the more or less educative schemes put forward by agents and professors in agriculture in agricultural education have been of help is due, perhaps, to the general brief experience and lack of pedagogical training on the part of teachers of agriculture. As a group these men are as well paid and as well prepared in the subject matter of their work as any group of secondary school teachers, probably better. But that some of them are really teaching boys instead of expounding agriculture is due to the use of organization and method under a home project scheme, evolved by others, rather than to their own initiative. That some few of them are doing work of excellent character is due to their initiative in modifying proposals that have merely helped others. In the hands of such men lies the solution of the problem of education for the farm boy. Their number can be increased by careful training in education of agricultural college students. With release of their energies by a more selective distribution of specific duties, an adequate allowance of time on a two or three period basis, and a tenure that shall include the months of the growing season, we may hope for closer adaptation of means to ends than now exists.

GOOD ORGANIZATION AND METHOD IN THE SECONDARY SCHOOL TEACHING AGRICULTURE CONSIST IN

I. Selection of problems according to
  A. Social factors of
    1. Local demand
    2. Demands of progress
    3. Type representation (scope)
    4. Continuity of expansion (correlation)
  B. Physical factors of
    1. Local opportunity
    2. Seasonal determination
  C. Psychological factors of
    1. Immediate activity
    2. Apperceptive basis in the pupil in terms of
      a. Native interests
      b. Age
      c. Farm, village, or urban experience
      d. School experience
      e. Present life
        1. In the community
        2. In the home
        3. In the school
        4. On the farm

II. Selection of the teacher in terms of
  1. Farm experience
  2. Technological training in Agriculture
  3. Teaching experience
  4. Pedagogical training
  5. Personality

III. Selection of equipment in terms of simplest adaptation to the problem
   1. Tools for energizing accomplishment
   2. Stock for productive use
      For comparative use
   3. Land for responsible production
   4. Books to answer questions
   5. Laboratory materials to answer questions

APPENDIX

EXTENT OF THE MOVEMENT TO TEACH AGRICULTURE IN SECONDARY SCHOOLS

The data on the following pages are not to be regarded as complete. They have been made up from various sources: Letters from state agents and superintendents, school directories, teachers' directories, and the digest of state laws published by the United States Bureau of Education. The compilation serves at least to show the wide distribution and the considerable number of schools engaged in the work.

State	High Schools		County Vocational Agriculture Schools	State Vocational Agriculture Schools	Special District Schools	Form of Special Aid
	No Special Aid	Special Aid				
Alabama	54				11	
Arkansas					4	
California	37					
Connecticut	5					
Florida	3					
Georgia					11	
Idaho	29					
Illinois	70					
Indiana	800[1]	4				⅔ maintenance
Iowa	525					
Kansas	441			1		
Maine		15				$500
Maryland		16				$400
Massachusetts		13	3			⅔ salary
Michigan	50		2			
Minnesota		176		3		$1000
Missouri	330					
Montana		3				$10 per pupil
Nebraska		23				½ maintenance to $1250
Nevada	0					
New Hampshire	25					
New Jersey	45					
New York		64		4		⅔ to ⅓ salary
North Dakota	43	5				$2020
North Carolina	15					
Oregon	6					
Oklahoma					6	
Pennsylvania		25				⅔ maintenance to $5000
Rhode Island	0					
South Dakota	26					
Texas		145				½ maintenance, $500–$1500
Vermont		11		2		All or ½ or ⅓ of $200
Virginia	11					
West Virginia	15					
Wisconsin	[2]		9			$250
Washington						
Total	*2530*	*500*	*14*	*10*	*32*	*12*

[1] 'About.' [2] No data.

TABLE I.—*Showing distribution of schools visited, by states*

Group Designation	Me.	N. H.	Vt.	Mass.	N. Y.	N. J.	Pa.	Md.	Ind.	Ill.	Wis.	Minn.	Ia.	Mo.	Ala.	Ga.
A. High Schools	2	3	2	3	9	0	2	1	2	2	1	4	1	1	3	3
A1. City High Schools					1				1	2		3		1		
A2. Town or District High Schools		1	1	3	8		2		1		1	1	1			
A3. County High Schools								1							2	
A4. Approved Academies	2	2														
A5. Congressional District															1	3
A6. Junior High School			1													
B. Agricultural Schools	0	0	1	2	1	1	1	0	0	0	3	1	0	0	0	0
B1. State			1		1							1				
B2. County				1							3					
B3. Philanthropic				1		1	1									
C. Preparatory				1												

TABLE II.—*Showing distribution of schools visited in which the work in Agriculture is supported wholly or in part by the state, by states*

Group Designation	Me.	N.H.	Vt.	Mass.	N.Y.	N.J.	Pa.	Md.	Ind.	Ill.	Wis.	Minn.	Ia.	Mo.	Ala.	Ga.
A. High Schools	2	0	0	3	9	0	2	1	1	0	1	4	0	0	3	3
A1. City High Schools					1							3				
A2. Town or District High Schools				3	8		2		1		1	1				
A3. County High Schools								1							2	
A4. Approved Academies	2															
A5. Congressional District															1	3
A6. Junior High School																
B. Agricultural Schools	0	0	1	2	1	0	1	0	0	0	3	1	0	0	0	0
B1. State			1	•	1							1				
B2. County				1							3				1	
B3. Philanthropic				1			1									
C. Preparatory																

TABLE III.—*Showing distribution of schools according to location*

Small cities, 7000–20,000 population
Towns, 1000–6000 population
Villages less than 1000 population

Group Designation	Large City	Subur-ban	Small City	Town	Village	Open Country
A. High Schools	1	4	5	15	9	5
A1. City High Schools	1	2	5			
A2. Town or District High Schools		2		12	5	
A3. County High Schools				1	1	1
A4. Approved Academies					2	2
A5. Congressional District				2		2
A6. Junior High Schools					1	
State Aid for Agriculture	1	3	1	9	4	2
B. Agricultural Schools	0	2	1	2	3	2
B1. State		1			2	
B2. County		1		1	1	1
B3. Philanthropic			1	1		1
C. Preparatory						1

TABLE IV.—*Showing expressed aims of Agricultural courses as given by principals and teachers*

Others aims expressed:
To reach the boy—two town High Schools
To dignify manual labor—Preparatory School

Group Designation	Vocational	Practical	Country Living	Prevocational	Liberalizing or Cultural	College Preparatory	Citizenship
A. High Schools	23	9	5	6	3	4	1
A1. City High Schools	4	2		3	1	3	1
A2. Town or District High Schools	16	4	1	1	1		
A3. County High Schools	1			2			
A4. Approved Academies	2	3			1	1	
A5. Congressional District			4				
A6. Junior High Schools		1					
Receiving State Aid for Agriculture	19	5	1	0	1	1	0
B. Agricultural Schools	10				1	1	2
B1. State	3				1		
B2. County	4					1	1
B3. Philanthropic	3						1
C. Preparatory School	1					1	

TABLE V.—*Showing ownership or possession of land*
TABLE V*a*.—*Showing ownership of live stock and farm machinery*
TABLE V*b*.—*Showing schools possessed of shop equipment*
TABLE V*c*.—*Showing schools of Group A in which Agriculture is housed in a separate building or in the basement, and schools having greenhouses*

	Table V		Table V*a*					Table V*b*			Table V*c*		
Group Designation	Owning Land	Hiring or Borrowing	Owning Farm	Neat Stock	Horses or Mules	Poultry	Machinery	Wood Shop	Wood and Forge	Farm Shop	Separate Building	Located in Basement	Green House
A. High Schools	13	6	7	7	5	9	10	12	10	5	6	8	
A1. City High Schools	4	1						3	2			4	2
A2. Town or District High Schools	6	5				3	3	7	3	3	5	3	1
A3. County High Schools	2						1	1	1				
A4. Approved Academies	1		3	3	1	2	2	1	1	1	1		
A5. Congressional District Schools			4	4	4	4	4		3	1			
A6. Junior High Schools												1	
B. Agricultural Schools			10	9	10	10	10		9	1			7
B1. State			3	3	3	3	3		3				2
B2. County			4	4	4	4	4		4				3
B3. Philanthropic			3	3	3	3	3		2	1			2
C. Preparatory			1	1	1	1	1			1			

TABLE VI.—*Showing relative amounts of agricultural laboratory material*
TABLE VI*a*.—*Showing possession of bulletins for reference*
TABLE VI*b*.—*Showing possession of volumes for reference in Agriculture*

	Table VI				Table VI*a*			Table VI*b*									
Group Designation	None	Scanty	Adequate	Complete	None	Scattered	Classified	0 0	1 9	10 19	20 29	30 39	40 49	50 74	75 99	100 149	150 more
A. High Schools	5	17	12	5	5	11	23	2	5	8	5	3	4	3	4	3	1
A1. City High Schools		4	4			5	3		1	3		2	1			1	1
A2. Town or District High Schools		10	5	4		3	16			4	2		3	3	4	2	1
A3. County High Schools	2			1	2		1	1	1			1					
A4. Approved Academies		1	3			1	3		1	1	2						
A5. Congressional District	3	1			2	2			3		1						
A6. Junior High Schools		1			1			1									
B. Agricultural Schools		2	2	6			10						1	1		1	7
B1. State			1	2			3										3
B2. County			1	3			4						1			1	2
B3. Philanthropic		2		1			3							1			2
C. Preparatory		1					1									1	

TABLE VII

Distribution of Titles of Reference Books

	Number Titles	*Regular Text*
Historical, Foreign Agriculture, Inspirational, etc.	15	0
Farm Plant Life	10	1
Elementary and General Agriculture	37	12
Agronomy	9	2
Soils and Manures	43	7
Farm Crops	44	5
Horticulture	19	3
Floriculture	12	0
Fruit Growing	37	4
Forestry	30	3
Insects and Diseases	34	4
Weeds	7	0
Animal Husbandry	24	5
Breeds and Breeding	26	4
Feeds and Feeding	16	6
Veterinary Science	18	2
Dairying	29	7
Poultry	35	5
Farm Management	19	6
Rural Social Conditions	29	0
Rural Sanitation	10	0
Farm Buildings	8	0
Rural Schools and Education	11	0
Agricultural Chemistry	9	2
Agricultural Physics	2	1
Special English	5	3
Farm Arithmetic	4	2
Farm Engineering, Mechanics and Shop	45	4
Farm Drawing	7	1
General Science	5	1

TABLE VIII—*Reference Books*

GENERAL AGRICULTURE

Author	Title	Publisher	Reference	Text	*Total*
Halligan	Fundamentals of Agriculture	Heath	7	2	*9*
Ferguson & Heath	Principles of Agriculture	Ferguson Publishing Co.	4	1	*5*
Voorhees	First Principles of Agriculture	Silver-Burdette	5	0	*5*
Jackson & Daugherty	Agriculture through the School and Home Garden	O. Judd Co.	7	0	*7*
Goff & Mayne	First Principles of Agriculture	American Book Co.	11	1	*12*
A. D. & E. W. Wilson	Agriculture for Young Folks	Webb	5	0	*5*
Bailey	Principles of Agriculture		10	0	*10*
Bailey	Am. Encyclopædia of Agriculture	Macmillan	10	0	*16*
Mayne & Hatch	High School Agriculture	American Book Co.	4	1	*5*
Waters	Essentials of Agriculture	Ginn	3	2	*5*
Warren	Elements of Agriculture	Macmillan	15	7	*22*
Brooks	Agriculture	Home Correspondence School	5	4	*9*
Wilkinson	Practical Agriculture	American Book Co.	5	1	*6*
Fisher and Cotton	Agriculture for Schools		5	0	*5*
Mann	Beginnings in Agriculture	Macmillan	13	6	*19*
Burkett, Stevens, and Hill	Agriculture for Beginners	Ginn	12	0	*12*
Call and Schafer	Laboratory Manual of Agriculture	Macmillan	4	1	*5*

AGRONOMY

Author	Title	Publisher	Reference	Text	*Total*
Clute	Agronomy	Ginn	9	0	*9*
Van Slyke	Fertilizers and Crops	O. Judd	11	2	*13*
Hunt and Burkett	Soils and Crops	O. Judd	10	3	*13*

SOILS

Author	Title	Publisher	Reference	Text	*Total*
Lyon and Fippin	Soils and Soil Management	Macmillan	15	2	*17*
King	The Soil	Macmillan	14	1	*15*
Roberts	The Fertility of the Land	Macmillan	10	0	*10*
Voorhees	Fertilizers	Macmillan	18	5	*23*
Fletcher	Soils	Doubleday-Page	2	5	*7*
Snyder	Soils and Fertilizers	Macmillan	12	2	*14*
Whitson and Walster	Soils and Soil Fertility	Webb	12	9	*21*
Hilgard	Soils	Macmillan	8	0	*8*
Vivian	First Principles of Soil Fertility	O. Judd	8	1	*9*
Hopkins	Soil Fertility and Permanent Agriculture	Ginn	15	0	*15*
Hall	The Soil	Dutton	6	0	*6*
Hall	Manures and Fertilizers	Dutton	6	0	*6*
Elliott	Practical Farm Drainage	Wiley	5	0	*5*

TABLE VIII—*Continued*

CROPS

Author	Title	Publisher	Reference	Text	*Total*
Bowman and Crossley	Corn		7	1	*8*
Sargent	The Corn Plants		6	0	*6*
Montgomery	The Corn Crops	Macmillan	7	0	*7*
Voorhees	Forage Crops	Macmillan	5	0	*5*
Livingston	Field Crops	Macmillan	12	0	*12*
Spillman	Grasses	O. Judd	8	0	*8*
Hunt	Cereals of America	O. Judd	15	2	*17*
Hunt	Forage and Fibre Crops	O. Judd	17	1	*18*
Wilson and Warburton	Farm Crops	Webb	11	17	*28*
Shaw	Clovers		5	0	*5*
Coburn	Book of Alfalfa	O. Judd	11	0	*11*
Fraser	The Potato	O. Judd	14	0	*14*
Grubb and Guilford	The Potato	Doubleday-Page	11	0	*11*
Shaw	Soiling Crops and the Silo	O. Judd	5	0	*5*
Dondlinger	The Book of Wheat	O. Judd	8	0	*8*
Myrick	The Book of Corn	O. Judd	5	0	*5*
Lyon and Montgomery	Examining and Grading Grains	Ginn	6	0	*6*

HORTICULTURE

Author	Title	Publisher	Reference	Text	*Total*
Bailey	The Farm and Garden Rule Book	Macmillan	8	0	*8*
Bailey	Plant Breeding	Macmillan	6	0	*6*
Taft	Greenhouse Construction	O. Judd	3	2	*5*
Bailey	American Encyclopædia of Horticulture	Macmillan	7	0	*7*
Green	Vegetable Gardening	Webb	6	2	*8*
Bailey	Principles of Vegetable Gardening	Macmillan	11	0	*11*
Lloyd	Productive Vegetable Gardening	Lippincott	9	6	*15*
Watts	Vegetable Gardening	O. Judd	6	2	*8*
Bailey	Manual of Gardening	Macmillan	5	1	*6*
Bailey	Garden Making	Macmillan	6	0	*6*
Beach	Apples of New York	New York State Experiment Station	10	0	*10*
Bailey	The Nursery Book	Macmillan	7	0	*7*
Sears	Productive Orcharding	Lippincott	12	9	*21*
Green	Popular Fruit Growing	Webb	15	9	*24*
Bailey	The Principles of Fruit Growing	Macmillan	19	1	*20*
Card	Bush Fruits	Macmillan	15	0	*15*
Waugh	Fruit, Harvesting and Marketing	O. Judd	9	0	*9*
Bailey	The Pruning Book	Macmillan	11	0	*11*
Waugh	The American Apple Orchard	O. Judd	10	0	*10*
Thomas	The American Fruit Culturist		5	0	*5*
Waugh	Systematic Pomology	O. Judd	5	0	*5*
Hedrick	Grapes of New York	New York State Experiment Station	6	0	*6*

TABLE VIII—*Continued*

FORESTRY

Author	Title	Publisher	Reference	Text	*Total*
Cheyney and Wentling	The Farm Woodlot	Macmillan	3	2	*5*

CONTROL OF PESTS

Author	Title	Publisher	Reference	Text	*Total*
Sanderson	Insect Pests		12	0	*12*
Smith	Economic Entomology	Lippincott	5	0	*5*
Chittenden	Insects Injurious to Vegetation		5	0	*5*
Comstock	A Manual of Insects		7	0	*7*
Weed	Farm Friends and Foes	Heath	6	0	*6*
Lodeman	The Spraying of Plants	Macmillan	7	0	*7*
Duggar	Fungous Diseases of Plants	Ginn	13	1	*14*
Georgia	A Manual of Weeds	Macmillan	7	0	*7*
	Farm Weeds	Ontario Department of Agriculture	5	0	*5*

ANIMAL HUSBANDRY

Author	Title	Publisher	Reference	Text	*Total*
Harper	Animal Husbandry for Schools	Macmillan	7	7	*14*
Harper	Manual of Farm Animals	Macmillan	8	1	*9*
Plumb	Beginnings in Animal Husbandry	Webb	8	10	*18*
Craig	Judging Live Stock		11	2	*13*
Day	Productive Swine Husbandry	Lippincott	8	0	*8*
Davenport	Principles of Breeding	Ginn	12	0	*12*
Marshall	The Breeding of Animals	Breeder's Gazette	3	3	*6*
Shaw	The Breeding of Animals	O. Judd	5	0	*5*
Plumb	Types and Breeds of Farm Animals	Ginn	16	5	*21*
Shaw	The Study of Breeds	O. Judd	7	1	*8*
Davenport	Domestic Animals and Plants	Ginn	3	2	*5*
Johnston	Book of the Horse		7	0	*7*
Coburn	Swine in America	O. Judd	8	0	*8*
Saunders	Short Horn Cattle		5	0	*5*
Roberts	The Horse	Macmillan	8	0	*8*
Gay	Productive Horse Husbandry	Lippincott	7	0	*7*
Henry	Feeds and Feeding	Author	17	3	*20*
Jordan	Feeding Animals	Macmillan	13	1	*14*
Woll	Productive Animal Feeding	Lippincott	4	2	*6*
Smith	Profitable Stock Feeding		3	5	*6*
Shaw	Management and Feeding of Cattle		5	0	*5*
	Diseases of the Horse	United States Bureau of Animal Industry	7	0	*7*
	Diseases of Cattle	United States Bureau of Animal Industry	6	0	*6*
Reynolds	Veterinary Studies	Macmillan	5	2	*7*
Mayo	Diseases of Animals	Macmillan	8	0	*8*

TABLE VIII—*Continued*

ANIMAL HUSBANDRY

Author	Title	Publisher	Reference	Text	*Total*
Eckles	Dairy Cattle	Macmillan	7	0	*7*
Corn	Dairy Bacteriology		14	0	*14*
Michels	Dairy Farming		4	2	*6*
Martin	Dairy Laboratory Guide		5	0	*5*
Van Slyke	Testing Milk and Its Products		6	1	*7*
Wing	Milk and Its Products	Macmillan	13	5	*18*
Farrington and Well	Milk Testing		9	3	*12*
McKay and Larsen	Buttermaking		3	2	*5*

POULTRY

Author	Title	Publisher	Reference	Text	*Total*
Watson	Farm Poultry	Macmillan	10	3	*13*
Robinson	Principles and Practice of Poultry Culture	Ginn	16	1	*17*
Lewis	Productive Poultry Husbandry	Lippincott	18	8	*26*
Brigham	Progressive Poultry Keeping		8	1	*9*
Valentine	How to Keep Hens for Profit	Macmillan	10	0	*10*
	American Standard of Perfection	American Poultry Association	7	0	*7*
Lewis	Poultry Laboratory Guide	Macmillan	6	0	*6*
Lippincott	Poultry Production		4	3	*7*

FARM MANAGEMENT

Author	Title	Publisher	Reference	Text	*Total*
Carver	Rural Economics	Ginn	5	1	*6*
Bexell and Nichols	Farm Accounting		5	1	*6*
Taylor	Introduction to Agricultural Economics	Macmillan	8	0	*8*
Card	Farm Management	Doubleday-Page	6	1	*9*
Warren	Farm Management	Macmillan	14	9	*23*
Boss	Farm Management		3	3	*6*
Warren and Livermore	Laboratory Manual of Farm Management	Macmillan	5	2	*7*
Hunt	How to Choose a Farm	Macmillan	6	0	*6*

FARM MECHANICS

Author	Title	Publisher	Reference	Text	*Total*
Davidson and Chase	Farm Machinery and Farm Motors	O. Judd	13	0	*13*
Davidson	Agricultural Engineering	Webb	7	6	*13*
Cobleigh	Handy Farm Devices		5	0	*5*
Brace and Mayne	Farm Shop Work		6	1	*7*

TABLE VIII—*Continued*

MISCELLANEOUS

Author	Title	Publisher	Reference	Text	*Total*
Snyder	Chemistry of Plant and Animal Life	Macmillan	6	2	*8*
King	The Physics of Agriculture		12	1	*13*
Roberts	The Farmers' Business Handbook	Macmillan	10	0	*10*
Bailey	The State and the Farm	Macmillan	5	0	*5*
Lipman	Bacteria and Country Life		7	2	*9*
Johnson	How Crops Grow	O. Judd	5	0	*5*
Osterhout	Experiments with Plants	Macmillan	5	0	*5*

TABLE IX.—*Showing office of teachers, length of time for which hired (tenure), whether farm or city born, farm experience of city teachers and teaching experience for forty-two teachers in Group A, thirty-five teachers in Group B, and two teachers in Group C*

Group Designation	Principal (and Agriculture)	Agriculture	Veterinary Science	Shop	Tenure Months			Farm Born	City Born	Farm Experience of City-born Teachers						Teaching Experience Years							
										Summers				Years									
					12	10	9			1	2	3	6	2	3	0	1	2	3	4	5	6–9	10 or more
A. High Schools	5	37			27	6	9	31	11	1	4	1	1	1	1	8	12	9	4		3	3	3
A1. City High Schools		8			6		2	4	4		2			1		1	3	2					2
A2. Town or District High Schools	1	20			17	2	2	18	3		2				1	4	6	5	3		2	1	
A3. County High Schools	2	2			2	1	1	3	1							1	1	1					1
A4. Approved Academies	1	3			1	2	1	1	3	1		1	1			1	2					1	
A5. Congressional District		4			1		3	4										1	1		1	1	
A6. Junior High Schools	1					1		1								1							
B. Agricultural Schools	9	18	2	6	30	5		32	3					1	1	4	4	1	4	2	5	7	8
B1. State	2	7	1	1	9	2		10	1						1	1		1	2		3	3	1
B2. County	4	7		4	13	2		14	1					1		3	2	1	1	1	1	4	2
B3. Philanthropic	3	4	1	1	8	1		8	1								1		1	1	1		5
C. Preparatory		2			2			2													2		

TABLE X.—*Showing training in pedagogy, degrees and subject matter training among seventy-seven teachers; and visitor's rating of sixty-two teachers*

A—Excellent. *B*—Good. *C*—Satisfactory. *D*—Poor. *E*—Failure.

Group Designation	Pedagogical Training					M.A.	A.B.	B.S.	Ph.B.	Ph.D.	Agriculture B.S.	Agriculture M.S.	Agricultural College		Graduate Student	Secondary School	D.V.S.	Practical	Technical	Grade				
	Summers		Years										Two years	Four years						A	B	C	D	E
	1	3	1	2	4																			
A. High Schools	3	1	12	1	1		3	2	1		34	1	1		1	2				1	9	19	5	1
A1. City High Schools	1	1	3				1		1		6	1									3	3	2	
A2. Town or District High Schools			6	1	1		1	1			19				1	1				1	5	9	2	
A3. County High Schools							1	1			2											1		
A4. Approved Academies	1		2								3		1								1	2		1
A5. Congressional District	1										3					1						3	1	
A6. Junior High Schools			1								1											1		
B. Agricultural Schools			3	4		1	1			3	18	6	2			1	2	5	2	1	12	11	2	
B1. State			1								8	2					1	1			4	3	2	
B2. County			1	4							7	1	2			1		2	2	1	4	6		
B3. Philanthropic			1			1	1			3	3	3					1	2			4	1		
C. Preparatory											1			1								1		

TABLE XI.—*Showing range of salaries paid to 406 teachers of Agriculture in Iowa high schools, receiving no aid from the state, salaries of 119 men teachers who have been students at agricultural colleges or universities of which agricultural colleges are a part, salaries of fifty-eight teachers of Agriculture, and salaries of fourteen women teachers attendant at agricultural colleges or universities of which agricultural colleges are a part.*

Computed from the Iowa State Educational Directory, 1915.

	All Teachers of Agriculture	Men Having Some Agricultural College Preparation	Women Teachers of Agriculture	Women Having Some Agricultural College Training
400– 449	2	1	0	0
450– 499	15	1	5	1
500– 549	8	2	5	0
550– 599	18	2	12	2
600– 649	14	2	7	0
650– 699	29	5	11	4
700– 749	24	5	6	2
750– 799	31	5	4	2
800– 849	34	8	1	0
850– 899	17	5	0	0
900– 949	61	16	4	1
950– 999	17	2	1	0
1000–1049	41	17	0	0
1050–1099	8	2	1	1
1100–1149	23	6	0	0
1150–1199	5	3	0	0
1200–1249	27	13	0	0
1250–1299	9	6	0	0
1300–1349	9	4	0	0
1350–1399	9	4	0	0
1400–1449	5	3	0	0
1450–1499				
1500–1549	4	3	0	0
1550–1599	0	0	0	0
1600–1649	5	4	1	1
1650–1699	1	0	0	0
1700	1	0	0	0

In the statement of preparation the length of agricultural college training is not given. A state university officer has informed the writer that such preparation in the majority of cases means only summer school attendance.

Salaries of teachers of Agriculture in state-aided schools of New York, 1915.

Number receiving	Academic Year	Summer	*Total*
1	$750	200	*$950*
2	800	200	*1000*
1	850	200	*1050*
11	900	200	*1100*
21	1000	200	*1200*
2	1050	200	*1250*
1	1075	200	*1275*
8	1100	200	*1300*
2	1150	200	*1350*
5	1200	200	*1400*
6	1300	200	*1500*
1	1400	200	*1600*
61			

Return of salaries of instructors in Agriculture for 136 state-aided high schools in Minnesota in fall of 1913. [By courtesy of Division of Agricultural Education, University of Minnesota.]

Number Receiving	Salary
6	$1000
11	1100
4	1150
43	1200
10	1250
15	1300
7	1300–1400
15	1400
3	1450
15	1500
6	1600–1850

Classification of fifty-nine teachers of Agriculture in state-aided high schools of New York, 1915, by colleges or schools in which agricultural training was received

New York State College of Agriculture	38
College of Agriculture, Syracuse University	3
Massachusetts Agricultural College	2
Michigan Agricultural College	2
Pennsylvania State College	1
Cortland Normal School (New York)	6
New York State School of Agriculture, Canton, New York	4
New York State School of Agriculture, Alfred, New York	3
	59

Number of schools in New York receiving state aid for Agriculture in

1911–1912	16
1912–1913	25
1913–1914	34
1914–1915	45
1915–1916	64

Number of high schools in Massachusetts receiving state aid for Agriculture:

1911	1
1912	2
1913	9
1914	10
1915	13

TABLE XII.—*Showing distribution of salaries of seventy-seven teachers*

Group Designation	$500 599	$600 699	$700 799	$800 899	$900 999	$1000 1099	$1100 1199	$1200 1299	$1300 1399	$1400 1499	$1500 1599	$1600 1699
A. High Schools	1	1	1	2	7	2	2	5	7	4	4	1
A1. City High Schools				1	1		1	3	1			
A2. Town or District High Schools				1	2	2	1	1	6	4	1	
A3. County High Schools		1						1			1	
A4. Approved Academies			1		1						1	
A5. Congressional District	1				2						1	
A6. Junior High Schools					1							
B. Agricultural Schools				1		2	2	2	2	1	4	3
B1. State						1					4	1
B2. County				1		1	2	2	2	1		
B3. Philanthropic												2
C. Preparatory												1

Group Designation	$1700 1799	$1800 1899	$2000 2099	$2100 2199	$2200 2299	$2300 2399	$2400 2499	$2600 2699	$2700 2799	$3000	$3150	$5000
A. High Schools	1	2	2	1								
A1. City High Schools	1	1										
A2. Town or District High Schools		1	1	1								
A3. County High Schools			1									
A4. Approved Academies												
A5. Congressional District												
A6. Junior High Schools												
B. Agricultural Schools	1	3	4		1	1		1	2	1	1	1
B1. State	1	1							1	1		
B2. County		2	2		1	1					1	
B3. Philanthropic			2					1	1			1
C. Preparatory							1					

TABLE XIII.—*Showing percentage of farm boys in agricultural courses*

Group Designation	0 9	10 19	20 29	30 39	40 49	50 59	60 69	70 79	80 89	90 100
A. High Schools	3	1	1	3	1	4	4	4	5	13
A1. City High Schools	3		1	2			1			1
A2. Town or District High Schools				1	1	4	3	3	3	4
A3. County High Schools		1								2
A4. Approved Academies								1		3
A5. Congressional District									2	2
A6. Junior High Schools										1
B. Agricultural	2		1			1		1	1	4
B1. State									1	2
B2. County			1			1				2
B3. Philanthropic	2							1		
C. Preparatory								1		

TABLE XIV.—*Showing ages of boys in agricultural courses in thirty-eight schools*

Group Designation	Over 14 Under 15	Over 15 Under 16	Over 16 Under 17	Over 17 Under 18	Over 18 Under 19	Over 19
A. High Schools	2	10	10	3	2	1
A1. City High Schools		2	4	1		
A2. Town or District High Schools	1	6	5			
A3. County High Schools		1				
A4. Approved Academies		1	1	2		
A5. Congressional District					2	1
A6. Junior High Schools	1					
B. Agricultural Schools			1	1	1	5
B1. State						3
B2. County				1	1	
B3. Philanthropic			1			2
C. Preparatory						1

TABLE XV.—*Showing units in Agriculture offered in schools of Group A*

Group Designation	½	1	1½	2	2½	3	3½	4	4½	5	6	8
A. High Schools	1	1	1	4	1	1	1	12	1	2	10	5
A1. City High Schools		1		2		1		3			1	
A2. Town or District High Schools				1	1[1]			3	1	1[1]	8	5
A3. County High Schools	1		1					1				
A4. Approved Academies								2		1	1	
A5. Congressional District							1	3				
A6. Junior High Schools				1								

TABLE XVI.—*Showing distribution of Group A, according to units of Agriculture offered and state aid*

Group Designation	No Aid [2]					State Aid for Agriculture							State Support			
	1	2	4	5	6	2	2½	3	4	4½	6	8	½	1½	3½	4
A. High Schools	1	2	4	2	1	1	1	1	5	1	9	5	1	1	1	3
A1. City High Schools	1	1	2			1		1	1		1					
A2. Town or District High Schools			2	1			1		1	1	8	5				
A3. County High Schools									1				1	1		
A4. Approved Academies				1	1				2							
A5. Congressional District															1	3
A6. Junior High Schools		1														

[1] Appears in High Schools and Vocational Departments.
[2] One Vermont school receiving only $66.67 from the state for Agriculture is included.

TABLE XVII.—*Showing number of schools in each group, length of regular courses to graduation, school year, state aid and boarding pupils*

Group Designation	Number	Years	Years	Years	Years	Weeks	Weeks	Weeks	Weeks	Weeks	Weeks	Weeks	Weeks	Weeks	Months	Receiving State Aid	Boarding Pupils
		4	3	2	1	24	32	34	36	37	38	39	40	45	12		
A. High Schools	39	39							28	1	4	3	3			29	9
A1. City High Schools	8	8							4		3	1				4	
A2. Town or District High Schools	19	19							15			1	3			16	
A3. County High Schools	3	3							2			1				3	1
A4. Approved Academies	4	4							2	1	1					2	4
A5. Congressional District	4	4							4							4	4
A6. Junior High Schools	1	1							1								
B. Agricultural Schools	10	3	4	5	2	1	1	1	2		1	1	1		2	9	9
B1. State Agriculture	3		1	2		1		1				1				3	3
B2. County	4	3	1	2	1		1		1		1		1			4	3
B3. Philanthropic Agriculture	3		2	1	1				1						2	2	3
C. Preparatory	1		1											1			1

TABLE XVIII.—*Showing distribution in Elementary Agriculture, Farm Crops, Horticulture*

First year	17	14	5	Third year	0	1	6	Fall	0	11	1	Spring	0	6	8
Second year	0	14	5	Fourth year	0	0	0	Winter	0	1	0	Full year	0	7	6

Group Designation	Elementary Agriculture			Farm Crops										Horticulture							
	Year	Units		Year				Season			Units			Year			Season		Units		
	1	1	½	1	2	3	4	F.	W.	S.	1	½	¼ or less	1	2	3	F.	S.	1	½	¼ or less
A. High Schools	12	6	6	9	10			8		3	7	10		3	1	6		6	2	7	1
A1. City High Schools	1		1	4				3			1	3		3				3		3	
A2. Town or District High Schools	3	1	2	2	3			2			3	2				2		1	1	1	
A3. County High Schools	2		2		1						1										
A4. Approved Academies	1	1		3	2			3		2	2	1			1	1		1	1	1	
A5. Congressional District	4	3	1		4					1		4				3		1		2	1
A6. Junior High Schools	1	1																			
B. Agricultural Schools	4	2	2	4	4	1		1	3	1	2	1		2	3		1	2			5
B1. State	1		1	3	2			1	2												
B2. County	1	1			2						1	1		1	2						3
B3. Philanthropic	2	1	1	1		1			1	1	1			1	1						2
C. Preparatory	1		1	1					1		1										

TABLE XIX.—*Showing distribution by years of Soils, Farm Crops, Animal Husbandry, Farm Management, Vegetable Gardening, Fruit Growing, Farm Mechanics, Dairying, Poultry, in seventeen proposed courses found in state manuals or other publications dealing with four-year courses in Agriculture for high schools.*

	Year	Year	Year	Year
Topic	1	2	3	4
Soils	5	4	1	4
Farm Crops	2	10	5	0
Animal Husbandry	0	5	9	1
Farm Management	0	0	0	13
Vegetables	5	4	2	1
Fruit Growing	1	4	6	3
Farm Mechanics	1	0	1	10
Dairying	0	4	10	1
Poultry	3	1	6	2

TABLE XX.—*Showing distribution by year, season, and units, of Agronomy, Soils, Fertilizers*

First year	5	12	2	Third year	1	2	3	Fall	2	5	2	Spring	1	5	5
Second year	10	4	2	Fourth year	0	4	3	Winter	1	2	1	Full year	12	10	0

Group Designation	Agronomy										Soils										Fertilizers									
	Year			Season			Units				Year				Season			Units			Year				Season			Units		
	1	2	3	F.	W.	S.	1½	1	½	⅓	1	2	3	4	F.	W.	S.	1	½	¼	1	2	3	4	F.	W.	S.	½	⅓	½ or less
A. High Schools	4	10		2			9	3	1	1	8		2	2	3		1	7	5		1		2	3	2		3	3	1	2
A1. City High Schools	1	1		1			1		1				2					2					1				1		1	
A2. Town or District High Schools	2	9		1			8	2		1	2				2	1		2	2				1	3	2		1	2		2
A3. County High Schools																					1						1	1		
A4. Approved Academies	1							1			2					2			2											
A5. Congressional District											4							1	3	1										
A6. Junior High Schools																														
B. Agricultural Schools	1		1		1	1		1	1		4	4		2	2	2	4		6	3	1	1	1			1	2	1		1
B1. State											1	2			1	1	1		1	2	1						1			1
B2. County	1		1		1	1		1	1		1			2		1	3		3	1										
B3. Philanthropic											2	2			1		1		2			1	1			1	1	1		
C. Preparatory																														

TABLE XXI.—*Showing distribution of time by topics in Agronomy and Animal Husbandry in state-aided Minnesota High Schools, 1915. By courtesy of Division of Agricultural Education, University of Minnesota*

Number of Weeks	Corn	Small Grains	Legumes	Root and Fiber Crops	Weeds	Grasses	Soils and Rotations	Horses	Dairy Cattle	Dairy-ing	Beef Cattle	Hogs	Sheep	Poultry	Nutrition	Average Number of Schools
1	1	2	1	15	15	7	8	3		3	6	5	18	14	7	7
2	2	3	24	42	33	46	16	10	13	12	17	16	37	34	18	21
3	8	7	25	18	33	25	15	12	14	20	14	21	24	17	7	18
4	18	19	37	23	22	21	18	23	27	19	33	31	20	22	22	23
5	5	3	8	6	2	6	7	16	10	8	6	10	1	3	6	6
6	10	19	12	4	9	7	11	25	28	13	19	19	3	4	14	13
7	2	5	1	1			3	1	3		2				2	1
8	18	25	5		3		9	11	2	9	4	1		3	6	6
9	4	8					2		1	1	1				1	
10	15	16	1				3	2	2	4					2	3
11	2									1						
12	17	8	1		1	1	5	2	1	1				1	3	3
14	1							2								
15	3						2									
16	7	2			1		1	1		1				1		
18	5					1	7	1		2			1		2	
20							1									
22	1															
24									1	1						
28							1									
30	1															
Total	*120*	*117*	*115*	*109*	*119*	*114*	*109*	*109*	*102*	*95*	*102*	*103*	*104*	*102*	*90*	

TABLE XXII.—*Showing distribution in Vegetable Gardening and Fruit Growing*

First year	10	1	Third year	2	18	Fall	1	18	Spring	14	4
Second year	7	5	Fourth year	2	5	Winter	0	1	Full year	7	7

Group Designation	Vegetable Gardening										Fruit Growing									
	Year				Season			Units			Year				Season			Units		
	1	2	3	4	F	W	S	1	½	¼ or less	1	2	3	4	F	W	S	1	½	¼ or less
A. High Schools	6	4	1	2			7	4	9			2	14		13		3	4	15	1
A1. City High Schools	1	1					1	1	1			1	1	2	1		1		2	
A2. Town or District High Schools	4	3		2			4	3	6				13	1	10		2	3	12	
A3. County High Schools														1	1					1
A4. Approved Academies			1				1		1						1				1	
A5. Congressional District	1						1		1			1						1		
A6. Junior High Schools																				
B. Agricultural Schools	4	3	1		1		6	1	4	3	1	3	3	1	4	1	1	3	4	1
B1. State	1	2					3		2	1	1	1	1		2	1			2	1
B2. County	2		1				2	1		2			2	1	1			2	1	
B3. Philanthropic	1	1			1		1		2			2			1		1	1	1	
C. Preparatory		1					1		1				1						1	

TABLE XXIII.—*Showing distribution by years and units of Animal Husbandry, Dairying, and Poultry*

First year	3	3	15	Third year	14	18	2
Second year	19	11	5	Fourth year	1	1	0

Group Designation	Animal Husbandry								Dairying								Poultry						
	Year				Units				Year				Units				Year			Units			
	1	2	3	4	1	½	⅓	¼ or less	1	2	3	4	1	½	⅓	¼ or less	1	2	3	1	¾	½	¼ or less
A. High Schools		18	13	1	9	22	1			7	17			2	2	1	12	2	1	3	9	2	1
A1. City High Schools		6	1		3	4				2	2			3	1		1				1		
A2. Town or District High Schools		5	10	1	3	12	1			3	12			14	1		10	2	1	3	8	2	
A3. County High Schools		1	2		2	1																	
A4. Approved Academies		4			1	3				1	2			3									
A5. Congressional District		2				2				1	1			1		1	1						1
A6. Junior High Schools																							
B. Agricultural Schools	3				2	1			3	4	1		4	2		2	3	3	1		3	1	4
B1. State	1	1			1												1	1	1				3
B2. County	2				1	1			2	2			4	1			1	1			2		
B3. Philanthropic										2	1			1		2	2	2			1	1	1
C. Preparatory			1		1							1	1										

TABLE XXIV.—*Showing distribution by years and units of Farm Management, Rural Economics, Farm Mechanics or Agricultural Engineering*

First year	0	1	2	Third year	6	0	7
Second year	3	0	5	Fourth year	27	5	15

Group Designation	Farm Management									Rural Economics						Farm Mechanics or Agricultural Engineering						
	Year			Units						Year			Units			Year				Units		
	2	3	4	1	¾	⅔	½	⅓	¼ or less	1	3	4	1	½	⅓	1	2	3	4	1	½	⅓
A. High Schools		2	24	12	1		10	1	1	1		3	1	3				5	14	7	11	
A1. City High Schools		1	4	2			2	1										1		1		
A2. Town or District High Schools		1	14	9			6			1		2	1	2				3	13	5	11	
A3. County High Schools			1		1																	
A4. Approved Academies			4	1			3					1		1				1	1	1		
A5. Congressional District			1						1													
A6. Junior High Schools																						
B. Agricultural Schools	3	3	3	1		1	3		4			1	1		1	2	5	2	1	4	1	4
B1. State	2	1					1		2							1	2	1		1		2
B2. County	1		3	1		1	1		1			2	1		1		2	1		3		1
B3. Philanthropic		2					1		1							1	1		1			1
C. Preparatory		1					1															

TABLE XXV.—*Showing distribution by years and units of Feeding, Breeds, Breeding, and Stock Judging*

First year	0	3	0	1	Third year	4	1	1	2
Second year	6	3	3	7	Fourth year	1	0	0	0

Group Designation	Feeding					Breeds					Breeding			Stock Judging			
	Year			Units		Year			Units		Year		Units	Year			Units
	2	3	4	½	¼ or less	1	2	3	½	¼ or less	2	3	¼ or less	1	2	3	¼ or less
A. High Schools		3		2	1		2		2						2		2
A1. City High Schools																	
A2. Town or District High Schools																	
A3. County High Schools																	
A4. Approved Academies																	
A5. Congressional District		3		2	1		2		2						2		2
A6. Junior High Schools																	
B. Agricultural Schools	6	1		3	4	3	1	1	4		3	1	4	1	5	2	6
B1. State	2			1	1	2	1		3		2		2	1	3	1	3
B2. County	3			1	2						1		1		2		2
B3. Philanthropic	1	1		1	1			1	1			1	1			1	1
C. Preparatory			1	1													

TABLE XXVI.—*Showing distribution in Ornamental and Landscape Gardening and Forestry*

First year	0	4	Fall	0	0
Second year	3	0	Winter	0	0
Third year	5	5	Spring	5	5
Fourth year	1	0	Full year	0	3

Group Designation	Ornamental and Landscape Gardening						Forestry					
	Year			Units			Year		Season	Units		
	2	3	4	1	½	¼ or less	1	3	S	1	½	¼ or less
A. High Schools	2				2		2	2	1	1	3	
A1. City High Schools												
A2. Town or District High Schools	2				2		2				2	
A3. County High Schools												
A4. Approved Academies								2	1	1	1	
A5. Congressional District												
A6. Junior High Schools												
B. Agricultural Schools	1	4	14	2	1	3	2	2	3	1		3
B1. State	1	1	2		1	1		1	1			1
B2. County		2	11	2	2	1	2		1	1		1
B3. Philanthropic		1	1					1	1			1
C. Preparatory		1	1		1			1	1			1

TABLE XXVII.—*Showing distribution by year and units of Carpentry, and Forgework or Blacksmithing*

First year	28	9	Third year	6	3
Second year	9	8	Fourth year	4	2

Group Designation	Carpentry									Not Differentiated Manual Training	Forgework or Blacksmithing						
	Year				Units						Year				Units		
	1	2	3	4	1½	1	¾	½	¼ or less		1	2	3	4	1	½	¼ or less
A. High Schools	21	5	3	3		3	6	9	5	5	8	3	2	2	1	4	4
A1. City High Schools	3	2				2	1	2		3	2				1	1	
A2. Town or District High Schools	11						5	4	2		3						3
A3. County High Schools	2								2	1							
A4. Approved Academies	2	1	1	1		1		1		1	1					1	
A5. Congressional District	3	2	2	2							2	3	2	2		2	1
A6. Junior High Schools																	
B. Agricultural Schools	7	4	3	1	1	2		1	3		1	5	1		1	3	2
B1. State	3	1						1	2		1	3				1	2
B2. County	3	2	1	1	1	1			1			2	1		1	2	
B3. Philanthropic	1	1	1		1	1											
C. Preparatory			1														

TABLE XXVIII.—*Showing length of periods, arrangement of periods*

Group Designation	Length of Periods in Minutes					Arrangement of Periods					
	30	40	45	50	60	Single Period	Four Single / One Double	Three Single / Two Double	Double Laboratory	All Triple	All Double
A. High Schools	1	31	5		2	8	5	7		3	16
A1. City High Schools		4	3		1		2	2			4
A2. Town or District High Schools		18	1			1	1	3		3	11
A3. County High Schools		2	1			2	1				
A4. Approved Academies		3			1	2		2			
A5. Congressional District		4				3	1				
A6. Junior High Schools	1										1
B. Agricultural Schools		6	3		1	2			6	2	
B1. State		1	2						3		
B2. County		4							3	1	
B3. Philanthropic		1	1		1	2				1	
C. Preparatory					1	1					

TABLE XXIX.—*Showing use of textbook; classroom method as recitation, lecture (major or minor method), topic assignments, supervised study, and the use of lesson plans by teachers of Agriculture; laboratory method, as use of manual, outlines, notebooks, individual or group experiments, demonstration by the teacher, no laboratory*

Group Designation	Text used	No text	Recitation	Lecture Major	Lecture Minor	Topic Assignments	Supervised Study	Lesson Plans Used	Manual	Outlines	Notes	Individual	Groups	Demonstration	No laboratory
A. High Schools	34	5	39	2	12	22	6	13	2	27	29	27	3	5	6
A1. City High Schools	8		8	1	2	5		1		6	6	6	1	1	1
A2. Town or District High Schools	15	4	19		6	15	4	11		18	18	16	2	1	
A3. County High Schools	3		3		1				1		1	1			2
A4. Approved Academies	3	1	4	1	1	2	1	1	1	2	3	3			1
A5. Congressional District	4		4		1					1	1	1		2	2
A6. Junior High School	1		1		1		1							1	
B. Agricultural Schools	10		10	6	4	5	1	1	3	6	9	9		1	
B1. State	3		3	3		2			3		3	3			
B2. County	4		4	2	2	1	1	1		4	4	4			
B3. Philanthropic	3		3	1	2	2				2	2	2		1	
C. Preparatory	1		1	1		1								1	

TABLE XXX.—*Showing shop method as correlated projects, process sequence, process to project, none; outdoor work as observation, practice, farm work, none; home projects, required, supervised, optional, unsupervised, prospective; school projects*

Group Designation	Shop				Outdoor Work				Home Projects					School Projects
	Correlated Projects	Process Sequence	Process to Project	None	Observation	Practice	Farm Work	None	Required	Supervised	Optional	Unsupervised	Prospective	
A. High Schools	11	8	7	13	12	18	6	3	15	12	7	10	5	10
A1. City High Schools		2	2	4	5	1		2	2	1	3	4	1	1
A2. Town or District High Schools	9	3	2	5	4	15			12	11	3	4	3	6
A3. County High Schools		1	1	1	1	1		1			1	1		1
A4. Approved Academies	1		1	2	1	1	2		1			1	1	1
A5. Congressional District	1	2	1				4							1
A6. Junior High Schools				1	1									
State Aid for Agriculture	8	3	5	7					13	12	5	6	4	9
B. Agricultural Schools	2	2	6		1		9		2	2				2
B1. State		1	2		1		2							
B2. County	1		3				4		1	1				1
B3. Philanthropic	1	1	1				3		1	1				1
C. Preparatory				1			1							

TABLE XXXI.—*Showing degree of correlation between classroom and outdoor work in Agriculture, and between classroom and laboratory work in Agriculture and showing correlation between Agriculture and other subjects in the curriculum*

Group Designation	Correlation of Classroom with Outdoor Work				Correlations Classroom and Laboratory		None	Botany	Zoology	Chemistry	Physics	Gen. Science	Arithmetic	English	Civics	Com. Geog.	Physiology
	None	Remote	Close	Remote in Time Only	Remote	Close											
A. High Schools	4	24	5	6	24	8	21	12	7	2	1	3	5	1			
A1. City High Schools	2	4	2		4	3	6	2									
A2. Town or District High Schools		10	3	6	15	4	9	8	6			1	2			1	
A3. County High Schools	2	1			1		2	1		1	1						
A4. Approved Academies		4			3		2	1	1			2	2				
A5. Congressional District		4			1		2			1				1			
A6. Junior High Schools		1				1							1				
State Aid for Agriculture	0	11	5	6	16	7	10	10	7	1	1	3	4	0		1	
B. Agricultural Schools		7	1	2	4	6	3	5	3	6	2		4	6	1		
B1. State		3[1]			3			2	1	2	1		1	2	1		1
B2. County		3[1]		1	1	3	2	2	1	2			1	2			
B3. Philanthropic		1	1[1]	1		3	1	1	1	2	1		2	2			
C. Preparatory		1				1	1										

[1] Showing close correlation in Horticulture in summer only.

TABLE XXXII.—*Showing adjustment of outdoor work to season and selection of agricultural subject matter in regard to demands of locality*

Group Designation	Season			Locality		
	None	Definite	Marked	Slight	Definite	Marked
A. High Schools	4	28	7	11	22	6
A1. City High Schools	3	4	1	3	3	2
A2. Town or District High Schools		13	6	2	13	4
A3. County High Schools	1	2		2	1	
A4. Approved Academies		4			4	
A5. Congressional District		4		3	1	
A6. Junior High Schools		1		1		
State Aid for Agriculture	0	16	6	1	16	6
B. Agricultural Schools	1	8	1	4	4	2
B1. State	1	2		2	1	
B2. County		4			3	1
B3. Philanthropic		2	1	2		1
C. Preparatory		1			1	

EXHIBITS OF LESSONS

LESSON IN FARM MECHANICS

Gas Engines

Eighteen boys worked in groups of three about six gas engines of different makes. Procedure consisted in taking apart the engine piece by piece, laying each carefully in a place assigned, then putting the engine together, starting and timing it. So the groups passed from engine to engine till all had been dismounted and set up again. Next the engines were taken apart, the pieces piled in indiscriminate heaps, and the round of reconstruction again accomplished. Finally, all engines were dismounted and the parts from all piled together. Then each group must sort out and replace the parts of its particular engine, and set it to running smoothly once more. The completion of the full routine occupied laboratory periods during several weeks.

LABORATORY EXERCISE IN FRUIT GROWING

1. Draw an outline of a round, oblate, conical, ovate, oblong apple.
2. Draw an acuminate cavity; an obtuse cavity.
3. Draw a cross section of an apple having core lines meeting.
4. Draw a cross section of an apple having core lines clasping.
5. Draw a cone shaped calyx tube.
6. Give definitions of the following:

   Calyx-lobes, skin-dots, closed core, open core, marginal stamens.

NOTE. Laboratory exercises in fruit growing are more likely to be concerned with the identification of varieties or practice in the processes of graft propagation, or the preparation of spraying mixtures, than in work like that above. Nevertheless, the insistence here indicated upon the botanical structure of fruits, as well as of other parts of the plant is not uncommon. The utilitarian aspect is not emphasized in such work.

LABORATORY LESSON IN FARM CROPS

High School

Every boy measured out ten cubic centimeters of oats from each of three samples, spread the samples upon clean paper, removed with a knife blade dirt, weed seeds, and chaff, and weighed again, to determine percentage of purity. During the sorting a smutty grain was discovered. This the teacher placed under a small microscope; then summoned the boys to examine it. After all had taken a look at the whole grain, he split it with a knife, laid the halves under the microscope, and again called for examination.

Finally he dictated the following questions, to be recorded and answered in the notebooks:

1. Of what value is a purity test?
2. Why must a purity test be made when you have grain for sale in this state?
3. Where can you get a purity test made?
4. Explain in full the value of this experiment to you.
5. Compare the three samples as to their content of weed seeds, chaff, etc. Which is most valuable and why?

A LESSON IN CORN JUDGING

High School

To every boy in the class was given a rack of ten ears of dent corn, and a printed score card. The boys spent the double period in scoring the individual ears, recording the score for each by its identification number. The following day the same racks were given out, and the boys, without the use of the score card or notebook, placed the ears in order of merit from right to left of the rack. When this placing had been completed notebooks were opened, placing of the previous day by score listed and the two sets of placings compared.

EXERCISE IN THE DAIRY LABORATORY OF A COUNTY AGRICULTURAL SCHOOL

Exercise 2. Milk Inspection

Object: To determine the amount of adulteration in a suspected sample of milk.

Method: There are two kinds of milk provided. One of the samples is chosen as a control sample, and the other has been adulterated by the addition of water. To find the amount of foreign water present, proceed as follows: Make a duplicate fat test of each sample. Take a lactometer reading of each sample and determine the S. N. F. present in each.

Then substitute into the following formula:—

$$100 - \frac{\text{S. N. F. of suspected sample} \times 100}{\text{S. N. F. legal standard}}$$

The suspected sample should be easily recognized; the previous work in dairying should tell you this.

NOTE. S. N. F. stands for Solids not Fat.

STUDENT'S PLAN FOR A BENCH HOOK

High School

1. Drawing of top view.
2. Drawing of end view.
3. Drawing of bottom cross piece.
4. Drawing of top cross piece.

Stock

1 piece white pine 7/8″ x 4½″ x 8½″
1 piece white pine 7/8″ x 1¾″ x 4½″
1 piece white pine 7/8″ x 1¾″ x 4″
4 screws 1½″, No. 8

Tools

Rule
Try square
Marking gauge
Saw
Plane
Brace and bit
Screw driver

Operations

1. Laying out
2. Surfacing one end
3. Planing one edge
4. Squaring one end
5. Laying out width, length, and thickness
6. Planing to thickness
7. Sawing to length
8. Planing to length
9. Sawing to width
10. Planing to width
11. Making blocks
12. Laying out screw holes
13. Boring screw holes
14. Assembling

LIST OF SHOP PROJECTS IN A COUNTY AGRICULTURAL SCHOOL

1. Bread board
2. Bench hook
3. Feeding trough for chicks
4. Bench
5. Milking stool
6. Window screen
7. Nail and staple box
8. Bird house
9. Folding bench
10. Fly trap
11. Chicken feed hopper
12. Clothes rack
13. Saw buck
14. Saw horse
15. Hive seat and tool box
16. Wagon pack
17. Roughage rack
18. Lawn seat
19. Fruit ladder or step ladder
20. King road drag
21. Planter
22. Stone boat
23. Land leveler
24. General purpose sled
25. Swing farm gate
26. Eveners and singletree
27. Two man weight carrier
28. Ladders
29. Fence and gate
30. Carpenters' tool box
31. Trapnest
32. Wheelbarrow
33. Carpenters' work bench
34. Chicken brooder coop and run
35. Wagon box
36. Ensilage rack for wagon
37. Wagon bed
38. Wagon box for wagon bed
39. Wagon box
40. Stock rack
41. Flat hay rack
42. Grain bed
43. Hay rack
44. The farm shop
45. Colony poultry house
46. Portable hog house
47. Pigeon house
48. Poultry house
49. Silo form
50. Corn crib
51. Implement shed
52. Hog house
53. Barn

LESSON IN BUILDING A POULTRY HOUSE

The floor of a forty-foot house had been laid, and the morning's task was in setting studding, laying plates, and cutting rafters. Two boys were assigned to laying off plates and studding, two to laying off and cutting rafters, and other groups, which did not rotate, to setting and spiking. The teacher stood aside, and did not interfere in any way. Several questions as to measurements he referred to the working drawings, which the class had prepared, and required verification by second measurement of all pencilings on the woodwork. So far did he refrain from suggestion or interference that rafter after rafter suffered from unskillful sawing, when a bit of direction and caution might have been useful.

The work went forward very busily and with little conflict of duties or delay.

A LESSON IN POULTRY

High School

Teacher. Choose one breed from each class of poultry and discuss it from the point of view of size, form and disposition.

Pupil made an excellent summary of the characteristics of the Light Brahmas, and a second began upon the Barred Rocks. He was interrupted by the obstreperous conduct of a white Wyandotte cock, tied up in a sack. The cock, rescued from the bag, was set up on the table, and criticized by the class as to form, color, and head points. Then the teacher listed the feathers by name, calling upon a boy to identify each group as it was named. When all had been so identified, every boy was called upon to show all classes of feathers in order, and to describe the distinctive features of each class. Finally the teacher showed the distribution of feathers in adjustment to muscular movements and protection.

LESSON IN FENCING A HEN YARD

High School

Teacher and janitor spent the entire forenoon in setting posts for an enclosure of ¾ acre of very uneven ground. In the afternoon for the first two periods the senior class grudgingly gave up the time 'for studying German for tomorrow's exam' to assisting in the wiring. To them the teacher gave occasional staccato orders but no instruction whatever. Two boys unrolled the wire, barking badly several young fruit trees in the process, and two were given hammer and staples. All corners were nailed by the janitor or the teacher. The attaching of the stretcher, a slow process, was performed by the janitor, the setting of braces by an interested neighbor, and the stretching itself by the janitor, the neighbor, and one boy. Two boys were ordered to stand on the wire to lower it, while the teacher drove the staples. The others ate apples and threw the cores at one another. At the end of the periods a group of second year students opened another roll of wire, and watched the janitor and teacher for eighty minutes.

LESSON IN THE USE OF A MANURE SPREADER

Philanthropic School

Sixteen boys armed with manure forks, pitch forks, shovels, and a coal scoop repaired, with the instructor, to a large heap of horse manure, up to which the farm foreman had backed a spreader. In two minutes the class had filled the body; whereupon the foreman drove off and spread the load. With the second load a boy was given the driver's seat, and instructed in the use of the feed lever. He then drove off, as had the foreman, chin on shoulder, in a meandering course across the field. No attention whatever was given to the matter of driving straight and turning properly. Every boy in his turn assumed so great a responsibility for the working of the gears that he could give little heed to his horses. The result was a ragged and uneven distribution of manure.

AN EXCURSION IN FORESTRY

High School

Before starting, boys were directed to fetch their notebooks, for, "There will be a report called for, and you must take notes." When all had gathered, the teacher spoke as follows:— "Be very observant. Take notes of the soil coverings, undergrowth, sod, litter of leaves, and see why it is difficult for seeds to take root. Pay attention to pine trees because deciduous trees are common here and evergreens are not. Note the bark, branching, cones and needles. Note differences of those in pines, spruces and cedars."

The excursion proved rather a wild frolic, some boys racing ahead, others dragging behind, doing 'stunts' and playing. The first stop was on a hillside where some cedars and arbor vitae grew. To the boys within hearing, the teacher pointed out the two species, leaving details of identification entirely to two or three of the faithful, who jotted down shape of needles in their books. About a mile further on a clump of white pines was found in the midst of birch and maple. Here the boys were somewhat interested in noting the differences of weight in dry limbs of pine and maple, in the presence of pitch in pine scars and in the large cones, but no mention was made of soil covering. A little further on some badly infected chestnut suckers were found. The teacher stopped to point out the appearance of chestnut bark disease, but most of the boys were too far away and too busy throwing stones to note the halt. The last halt was made under spruces planted in a farm yard. The boys collected fir cones but secured no needles because the yard was carefully raked and the trees trimmed high. On arrival at the school, the boys gathered about the faithful and copied their notes.

A PROBLEM LESSON

The pupils of a special school had planned a husking bee. To the class in Farm Crops, quite appropriately, the preparation for the husking was assigned. At the time of the visitor's arrival, the class was on its way to the machine shed in which the 'party' was to take place.

The next move was to run out wagons and machinery. This done, the class gathered in one end of the shed for making of plans. Initiative remained with the boys for the most part. The first question raised was as to the number who would attend the husking, the next as to accommodating the number in the shed. The first was readily answered, the second called for thinking. Two boys paced off the dimensions of floor space to meet the first inquiry, then two sat down back to back while the others estimated the space they occupied. Then the problem became one of arrangement of the rows of corn in the shuck; should they run in short rows parallel with the end of the rectangular shed or lengthwise, with a maze formation. Considerable argument ensued before choice of the longitudinal arrangement prevailed, as that which provided the greater elbow room.

Next the teacher asked, "Now what is the best way to get that corn in?" One boy suggested that, since the distance to shocks in the field was short, all get together after school and bring it in in wheelbarrows. But to this the majority demurred on the ground that they must be at home for work in the afternoon. Everyone volunteered to pay his share toward pulling ears and placing them in the shed, and it was voted to call on boarding pupils who were earning their way to do the work. Then came the question, How shall they be paid? Some suggested payment by the hour, others by amount of work done. The 'piece plan' prevailed on a vote, the bushel being chosen as a unit.

The teacher's second question was: "How much are you going to pay a bushel?" Various prices, from two to ten cents, were suggested, nobody knew what a fair price might be. "How can we find out?" came as the third question. Out of this arose several vague suggestions, each of which was rejected as no better than guess work, till a boy put forth the following: "I guess we fellows are worth about ten cents an hour when we work our best. That is what most of us get when we work for pay. Now, if we all go out to the corn rows and pull corn and lug it in for half an hour, we can figure out what is a fair price per bushel, by the result." On this plan class and instructor were hard at work, when the visitor left for another class meeting.

NOTE. Against this lesson may be advanced the criticism that the knowledge acquired was hardly worth the effort involved. In favor of it, the criticism that the method involved is that of the real problem. The whole experience moved naturally in thought and action to the solution of no fictitious problem. It was in very marked contrast with the usual haphazard or rule of thumb procedure in outdoor work.

MASSACHUSETTS PROJECTS

Vegetable Garden	123	Flowers	2
Poultry	95	Calves	2
Farm Crops	83	Steers	1
Orchard	48	Sheep	1
Dairy Cattle	18	Bees	1
Swine	10	Shrubs	1
Small Fruit	6		
			391

NEW YORK PROJECTS

Poultry	352	Bush and Small Fruits	53
Potatoes	126	Dairy Cattle	39
Corn	62	Other Farm Animals	15
Garden	81	Miscellaneous	78
Apples	30		
Other Tree Fruits	26		
			862

Means of transportation to home projects from 42 state-aided schools of New York, 1915.

Automobile	20	Livery	6
Bicycle	10	Railroad	2
Motorcycle	3	Boat	1
Horse (owned)	4		

Transportation expenses.

No. receiving:	1	$5.	No. receiving: 9	$50.
	1	10.	3	75.
	2	15.	4	100.
	3	25.	1	125.
	2	40.		

Board of Education furnishes:

Automobile	4	Bicycle	1
Horse	2	Gasoline	3

Plan of visits to projects at Concord High School.

Summer: twice a week.
Fall and Spring: as often as possible.
Winter: once a month.

Record of visits to nineteen home projects of pupils.

No. of Project	*No. of Visits*	*No. of Project*	*No. of Visits*
1	22	11	25
2	36	12	31
3	21	13	17
4	27	14	25
5	25	15	26
6	27	16	20
7	30	17	23
8	28	18	27
9	26	19	33
10	27		

REPORT OF EXTENSION WORK OF TEACHERS IN THE ALBERT LEA HIGH SCHOOL

Quoted from letter of Mr. W. E. Hedgecock, in charge

"The Extension Work has been so interwoven from one year to another that it will be rather difficult to separate last year from the previous years and for that reason I will give you the work as conducted during the past two years, as Mr. Hibbs has been with me both years.

"The Extension Work was divided into two main projects. The introduction of alfalfa into the county and better dairy farming.

The alfalfa acreage has been increased from twelve acres in 1913 to almost 300 acres in 1915, and I expect to hear that at least 600 acres were cut during the spring and summer of 1916. The Dairy Industry has been promoted through the Pioneer Cow Testing Association, which numbered 700 cows at the completion of the reorganization in June, 1915. This Association included thirty-five herds of cattle.

"All the farmers with the exception of a very few have taken to weighing their milk night and morning, keeping a record of it on sheets furnished by the United States Department and distributed by the Agricultural Department of the High School. The men with one or two exceptions have erected silos within the past few years and a large number during the last two years. They have all improved in the feeding of their dairy cows and at the present time we have one Guernsey Breeding Association a year and one half old.

"The remainder of our time was devoted to various lines of work such as advice in building, remodeling a dairy barn, making plans for machine sheds, arranging for dairy meetings and organizing and aiding the farmers' clubs of the County. The County at the time I left had thirteen farmers' clubs holding meetings once a month. We also aided in conducting a one week farmers' short course cooperating with the State Extension Division. We also held each year a three months' short course for farmer boys and girls, fifteen years of age and over. This last year we had twenty-five pupils enrolled whose age would average probably between eighteen and nineteen years.

"We also cooperated with the Farm Management Department of the College of Agriculture in obtaining the consent of two men to keep records of farm work and both horse and man labor. We also cooperated in promoting the boys' and girls' clubs throughout the County."

SUMMARY OF EXTENSION WORK AT THE CONCORD, MASS., HIGH SCHOOL

From the Report of Mr. A. W. Doolittle, in charge

Fumigated six greenhouses with Cyanide for White Fly. Conducted two experiments on spraying cucumbers in greenhouses for Red Spider.

Conducted Apple Packing School January 4 to 9. Gave practical work in box, barrel, and fancy packages. Enrolment twelve adults.

Supervised and arranged for Extension School by the Massachusetts Agricultural College for five days in February. Total attendance of 600 men and 400 women. Concord Library issued catalogs of Agriculture and Domestic Science books, which were mailed and distributed to farmers and housewives.

Most of the spare time for a month was given to promoting this school. Meetings were held, letters written, programs distributed and personal work done. Mr. Bevan visited sixty-five farms to urge attendance. He also took a census of fifty farms.

An Agricultural census of the town was started and nearly completed; a special survey on asparagus and strawberry growing made. A total of 1606 calls were made on the telephone.

Assisted with Concord Grange Fair held in October. Helped in advertising and distribution of premium lists. Aided in judging.

Judged fruit and vegetables at Framingham Fair, September 22, and at Home Garden Exhibit, Everett, September 16.

Took active interest in local Home Garden Association. Headed committee on premium list and exhibits. Distributed seed and fertilizers. Assisted in managing the two exhibits, receiving, displays, judging, etc. Three hundred boys and girls took part.

Acted as official milk tester for the Board of Health for which apparatus valued at $175.00 has been given. Results of tests are published in local papers. 108 samples have been reported on.

Gave four lectures on Agricultural subjects in Concord, Bedford, Sudbury, and Dunstable. Total attendance 175.

Aided in Clean-up Week campaign.

Made drawings to scale of school plant in preparation for ornamental planting.

Gave public demonstration of milk testing for Board of Health.

Wrote fifteen Agricultural articles for local papers.